ELECTROANALYTICAL TECHNIQUES IN CLINICAL CHEMISTRY AND LABORATORY MEDICINE

ELECTROANALYTICAL TECHNIQUES IN CLINICAL CHEMISTRY AND LABORATORY MEDICINE

Joseph Wang

New Mexico State University

VCH

Joseph Wang
Department of Chemistry
New Mexico State University
Las Cruces, New Mexico 88003

616.075
W184e

Library of Congress Cataloging-in-Publication Data

Wang, Joseph, 1948–
    Electroanalytical techniques in clinical chemistry and laboratory
medicine / Joseph Wang
      p. cm.
    Includes index.
    ISBN 0-89573-342-0
    1. Electrochemical analysis.   2. Electrodes, Ion selective.
3. Biosensors.   4. Chemistry, Clinical.   I. Title.
    RB41.W36   1988
    616.07'5—dc 19

88-20580
CIP

British Library Cataloguing in Publication Data

Wang, Joseph, 1948–
    Electroanalytical techniques in clinical chemistry and laboratory medicine.
    1. Medicine. Laboratory techniques.—Manuals
    I. Title
    610:28
    ISBN 0-89573-342-0

Printed in the United States of America.

ISBN 0-89573-342-0 VCH Publishers
ISBN 3-527-26799-9 VCH Verlagsgesellschaft

Distributed in North America by:

VCH Publishers, Inc.
220 East 23rd Street
Suite 909
New York, New York 10010

Distributed Worldwide by:

VCH Verlagsgesellschaft mbH
P.O. Box 1260/1280
D-6940 Weinheim
Federal Republic of Germany

PREFACE

Over the last two decades, there has been substantial progress in the discovery of substances used in the diagnosis, prevention, and treatment of human diseases. These developments have resulted in increased demands for sensitive and specific analytical methods. Many analytical techniques are required to solve analytical-clinical problems. The instrumental methods for quantitation which are most commonly used in a clinical laboratory fall into four basic categories: spectrophotometric, chromatographic, radiometric, and electrochemical analysis. The choice of the appropriate analytical technique to solve a specific problem is often dictated by the analyte, its concentration level, and the sample matrix.

Electroanalysis is a major branch of analytical chemistry that makes use of electrical measurements for analytical purposes. Electroanalytical techniques have been found to be quite widely applicable to many problems of clinical and pharmaceutical interest. Dissolved oxygen can be continuously monitored in body fluids with amperometric membrane electrodes. Ion-selective electrodes permit reliable and rapid measurements of potassium, sodium, lithium, calcium, and hydrogen ions. The excellent detection limits that can be achieved with liquid chromatography/electrochemistry, pulse voltammetry, and electrochemical immunoassays show great promise for the determination of therapeutic drugs in blood serum. Electrochemical biosensors are being used for the measurement of glucose, urea, and creatinine. Trace levels of toxic metals, such as lead and cadmium, have been determined in biological matrices using stripping voltammetry. Electroanalytical methodologies are especially suitable when small sample sizes, in-vivo analysis, or detection in flowing streams is concerned. These and other properties and capabilities of electroanalytical techniques are certain to have an increasing impact on clinical chemistry, particularly as analytical devices move from the central laboratory to the bedside, to the physician's office, and on to the home.

This book attempts to introduce the basic principles of electroanalytical techniques and to show how they are used in the clinical field. Detailed theory is not given, and the approach is primarily experimental. Numerous examples are used to illustrate the scope and possibilities of electroanalysis in the clinical laboratory. While references are selective, I hope that the more than 700 references that are cited will offer adequate entry to the literature. The material is divided into five chapters. Chapter 1 is devoted to finite current controlled-potential techniques. Ion-selective electrodes are covered in Chapter 2. Chapter 3 deals with electrochemical biosensors, including enzyme electrodes and immunosensors, while electrochemical detectors for flowing streams (e.g., liquid chromatography, automated analyzers) are discussed in Chapter 4. In-vivo electrochemistry is the topic of Chapter 5.

Finally, I wish to acknowledge my wife, Ruth, and my daughter, Sharon, for their love and patience; the editor, Professor Gary Christian, for many, many useful suggestions; the fine people at VCH Publishers and Barbara O. Chernow and Douglas L. Bell at Chernow Editorial Services for their great cooperation and assistance; and Mr. Vince Villa, who did much of the typing.
Thank you all!

Joseph Wang
October 1987

CONTENTS

CHAPTER 3

Electrochemical Biosensors   *79*

CHAPTER 1

Voltammetry and Other Controlled-Potential Techniques

1-1 INTRODUCTION

Voltammetry is a powerful and versatile analytical technique that offers high sensitivity, precision, and accuracy as well as a wide linear range, with relatively low-cost instrumentation. Although voltammetry represents a rather specialized area of instrumental analysis, it is a powerful tool in the armory of the clinical chemist. Voltammetric measurements of numerous electroactive species of biological significance (drugs, hormones, vitamins, and metals) have been reported. In addition, voltammetric procedures have been used extensively to elucidate the redox properties of important biological compounds such as pyridine coenzymes and flavin derivatives.

Voltammetry represents a class of electroanalytical methods in which the current at the working (polarized) electrode is measured as a function of the potential applied to that electrode. The term *working electrode* is reserved for the electrode at which the reaction of interest occurs. The potential of the electrode serves as the driving force for the electrochemical reaction; i.e., it is the controlled parameter that causes the chemical species present in the solution to be electrolyzed (reduced or oxidized) at the surface. Thus, it can be viewed as a measure of the electron pressure. The reducing or oxidizing strength of the electrode is controlled by the applied potential. For example, as the potential of the electrode becomes more negative, the electrode becomes a stronger reductant (electron source). More positive potentials will favor oxidations. In a voltammetric experiment, the potential is varied in some systematic manner, e.g., linear ramp or a pulse train. If an electroactive species (molecule or ion) is present, a current will be recorded when the applied potential becomes sufficiently negative or positive for it to be electrolyzed (on both sides of the standard potential E° for the redox reaction of the species). For systems controlled by the laws of thermodynamics, the applied potential E controls the concentration of the electroactive species:

$$o + n\,e^- \rightleftharpoons R \tag{1-1}$$

1

at the electrode surface (at 25°C) according to the Nernst equation:

$$E = E° + 0.059/n \log C_O/C_R \qquad (1\text{-}2)$$

where C_O and C_R are the concentrations of the oxidized and reduced forms of the electroactive species (O and R, respectively), and n is the number of electrons transferred in the reaction. On the positive side of $E°$, the oxidized form of the redox couple is stable, whereas the reduced form tends to undergo oxidation if it reaches the electrode. The current resulting from a change in oxidation state of the electroactive species is termed the *faradaic current* because it obeys Faraday's law. [According to this law, the electric charge, coulombs (C), involved in a redox reaction of 1 mole of substance is equivalent to $n \times 96485$ C.] The faradaic current is a direct measure of the rate of the redox reaction taking place at the electrode. This depends mostly upon two things: (1) the rate at which the species moves from the bulk of the solution to the electrode (this process is called *mass transport*), and (2) the rate at which electrons transfer from electrode to solution species and vice versa (this process is called *charge transfer*). This picture is oversimplified; a complete description of the redox reaction may involve additional processes, such as surface or chemical reactions. The process with the slower rate controls the magnitude of the current. The terms *reversible, irreversible,* etc., are used to describe such control, and thus depend on the magnitude of the heterogeneous rate constant for the electron-transfer process compared with the average rate of mass transport. A complete understanding of electrode processes requires knowledge of kinetics, thermodynamics, hydrodynamics, solution and surface processes, and basic electrochemical principles. A comprehensive treatment of all these topics is beyond the scope of this discussion. Several monographs [1-4] are suggested for this purpose.

The resulting current-potential plot is called a *voltammogram*. The voltammo-gram, the electrochemical equivalent of the spectrum obtained in spectroscopy, is a display of current (vertical axis) versus potential (horizontal axis). The current can be considered as the response signal to the potential excitation waveform. Depending on the nature of the measurement, a wave- or peak-shaped response is observed. The exact shape of the response is governed by the processes involved in the electrode reaction. For mass-transport-controlled (reversible) reactions, the current i is related to the flux of material to the surface, as described by the following equation:

$$i = nFAD_O \, (\delta C_O/\delta x)_{x=0,t} \qquad (1\text{-}3)$$

where F is the value of the Faraday, A is the electrode surface area, D_o is the diffusion coefficient of O, and $(\delta C_O/\delta x)_{x=0,t}$ is the slope of the concentration-distance profile at the electrode surface. For a mixture of electroactive species, the voltammogram is the summation of the waves or peaks for the individual components. In addition to the faradaic current, a background (residual) current— unrelated to the redox reaction of interest—flows through the cell; this current limits detectability and dictates the mode of measurement.

By careful interpretation of the voltammogram, important analytical information (quantitative and qualitative) is obtainable. The potential at the rising portion of the wave, or at the peak, is related to $E°$ for the couple and therefore provides qualitative information. The faradaic current is proportional to the concentration of the electroactive species, with the detection limit being determined by the relative size of this current compared with the background current. The most advanced voltammetric procedures, aimed at reducing the contribution of the background current relative to the measured current, have detection limits at the nanomolar and subnanomolar levels. The improved sensitivity of voltammetric procedures has enabled their application to the assay of low concentrations of species of biological and pharmaceutical interest in body fluids and tissues. Essential to the growing interest in voltammetric measurements in clinical medicine has been the introduction of commercial multipurpose voltammetric instruments. The following sections describe the basic principles of various voltammetric techniques and their applications in analyses of biological samples. Detailed theory is not given; the approach is primarily experimental.

1-2 INSTRUMENTATION

Voltammetric measurements are carried in an electrochemical cell, which is usually a covered beaker of 5- to 50-ml volume. The cell contains three electrodes (working, reference, and auxiliary), which are immersed in the sample solution (Figure 1-1). For most reductions, the working electrode is a dropping mercury electrode (DME) or a hanging mercury drop electrode (HMDE). A convenient design of such mercury electrodes is available in the form of the so-called static mercury electrode. Solid electrode voltammetry, which is typically employed for oxidative work, can use stationary or rotating electrodes, usually in a disk configuration. Such electrodes consist of a circle of the electrode material, sealed in a tube of an insulating material (e.g., Teflon, glass). The rotating disk electrode, in particular, provides a very versatile tool for voltammetric measurement of oxidizable compounds, because its hydrodynamics are rigorously understood and are advantageous. Electrodes constructed from carbonaceous materials (e.g., carbon paste, glassy carbon) exhibit relatively low background currents and are useful for numerous electroanalytical applications. For certain applications, metallic electrodes made of platinum or gold are useful. Unlike the renewable surface of the DME, solid electrodes may exhibit a gradual decrease in surface activity due to adsorption of surfactants (present in biological samples) or accumulation of reaction products. Expertise in handling solid electrodes, including various pretreatment or reactivation procedures, is now well established. For example, different treatments based on mechanical polishing, electrochemical oxidation, high vacuum, or laser irradiation have been developed to assure reproducible data at the commonly used glassy carbon electrodes. In addition to the ordinary electrodes described above, new types of working electrodes based on surface modification and composite materials are expected to find increasing use in the near future. For example, protein

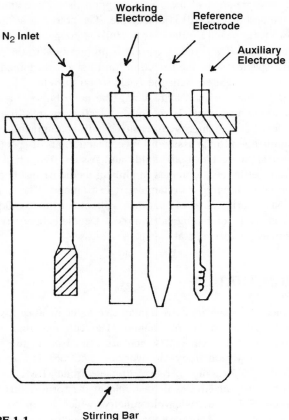

FIGURE 1-1
Schematic diagram of a cell for voltammetric measurements.

adsorption may be minimized using electrodes coated with an appropriate permselective polymeric film. Electrodes modified with an appropriate catalyst can substantially enhance the electron-transfer properties of numerous biological compounds. In addition, biocomponents (particularly enzymes and antibodies) can be attached to the surface, as desired for various biosensing applications (see Chapter 3). Another type of working electrode, the ultramicroelectrode, has received increased attention in recent years. This is due to unique, and attractive, properties such as nonlinear planar diffusion (which results in large current densities and independence of solution movement), fast response time, low ohmic drop, and suitability to analysis of very small sample volumes, as well as to in-vivo measurements.

The reference electrode provides a stable potential against which the potential of the working electrode is compared. Such "buffering" against potential changes is commonly achieved using the silver–silver chloride and saturated calomel reference electrodes. An inert conducting material, such as platinum wire or graphite rod, is usually used as the current-carrying auxiliary electrode. The three electrodes, as well as the tube used for bubbling the deoxygenation gas (used in reductive work since oxygen is electroreducible), are supported in four holes in the cell cover. Various micro cells with $0.5-2$-ml and $20-100$-μl volumes have been used in conjunction with mercury drop and solid electrodes, respectively, as required in many clinical situations.[5-7] The major piece of equipment required for controlling the potential of the working electrode while monitoring the resulting current is a voltammetric (polarographic) analyzer. The main characteristic of the modern voltammetric analyzer is the potentiostatic control of the working electrode potential. The circuit is arranged so that potential control is maintained between the working and reference electrodes, while the current passes between the working and auxiliary electrodes. No current passes through the reference electrode.

Multipurpose microprocessor-based voltammetric analyzers are now commercially available from various sources such as EG&G Princeton Applied Research, Bioanalytical Systems Inc., Sargent-Welch, Brinkmann (Metrohm) Instruments, Amel, and Tacussel. (Electrodes and cells can also be obtained from these sources.) Semiautomation or full automation (including additions of reagents) can be achieved with the new generation of microprocessor-based instruments.

1-3 SAMPLE CONSIDERATIONS

The undesired migration movement of charged analytes under the influence of the electrical field is reduced to a negligible value by the presence of high concentrations ($0.05-0.5$ M) of supporting electrolyte, for example potassium chloride or hydrochloric acid. The supporting electrolyte also decreases the ohmic resistance of the solution and thus insures the conductive media that voltammetry requires.

Buffer systems, such as acetate-acetic acid, are used as supporting electrolytes when pH control is essential. Tetraalkyl ammonium salts are often used as supporting electrolytes in conjunction with nonaqueous solvents. (The use of such solvents is frequently required for organic analytes that are barely soluble in water.)

Because the reduction of oxygen yields a background response over a large portion of the cathodic potential range, it is necessary to remove dissolved oxygen from the solution prior to the measurement of reducible species. This is accomplished by bubbling nitrogen or helium through the solution for several minutes. To avoid oxygen intrusion, the cell should be blanketed with the gas while the voltammogram is being recorded.

1-4 POLAROGRAPHY

Polarography is a form of voltammetry in which the working electrode is dropping mercury. Because of the special properties of this electrode, polarography has been widely used for the determination of many reducible species of biological significance. The technique was invented by Heyrovsky in Czechoslovakia in 1922, who received a Nobel prize for this work.

The DME consists of a 15-cm long thin-bore capillary connected by a flexible tube to an elevated reservoir of mercury (Figure 1-2). By adjusting the height of the mercury column, one may vary the drop time; drop times ranging from 3 to 6s are most common.

One advantage of the DME is the high hydrogen overvoltage that extends the cathodic potential range. This feature makes accessible to polarography many analytes with negative redox potentials. Another advantage is that the surface of the DME is continually being renewed, minimizing the effects of being fouled or poisoned. One limitation of the DME is the ease with which mercury is oxidized. Thus, measurements are largely restricted to substances reducible within the potential range of the DME, approximately $+0.3$ to -2.0 V (versus the saturated calomel electrode), depending on the solvent, electrolyte, or pH.

The excitation signal used in conventional polarography is a linearly increasing potential ramp. For a reduction, the initial potential is selected so that the reaction of interest does not take place. The potential is then scanned cathodically while the

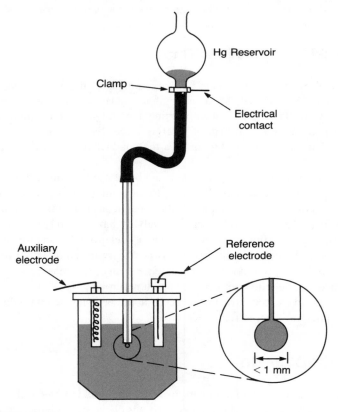

FIGURE 1-2
The dropping mercury electrode.

current is measured. At a sufficiently negative potential, reduction of the analyte commences, and the current rapidly increases to its limiting (diffusion-controlled) value. At this plateau region, any analyte particle that arrives at the electrode surface by diffusion instantaneously undergoes an electron-transfer reaction. The resulting polarographic wave is shown in Figure 1-3.

The potential where the current is one-half of its limiting value is called the *half wave potential*, $E_{1/2}$. The half-wave potential (for electrochemically reversible couples) is related to the formal potential, $E°$, of the electroactive species according to

$$E_{1/2} = E° + 0.059/n \, \log (D_R/D_O)^{1/2} \qquad (1\text{-}4)$$

where D_R and D_O are the diffusion coefficients of the reduced and oxidized forms of the electroactive species, respectively. Because of the similarity in the diffusion coefficients, the half-wave potential is usually similar to the formal potential. Thus, the half-wave potential, which is a characteristic of a particular spe-

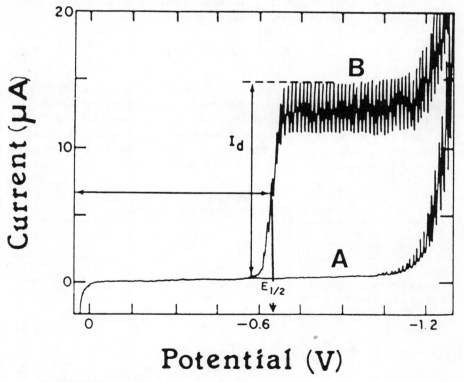

FIGURE 1-3
Polarograms for (A) 1 M hydrochloric acid and (B) 4×10^{-4} M
Cd^{+2} in 1 M hydrochloric acid. $E_{1/2}$ is the half-wave potential, while
i_d represents the limiting current.

cies in a given supporting electrolyte solution, is independent of the concentration of that species. Therefore, by measuring the half-wave potential, one can identify the species responsible for an unknown polarographic wave. Extensive tables of half-wave potentials exist in the literature[8] and can be used for this purpose. Values of half-wave potentials for several reducible organic functionalities, common in biological compounds, are given in Table 1-1. Many redox reactions of such functionalities involve hydrogen ions. Such reactions can be represented as

$$R + nH^+ + ne^- \rightleftharpoons RH \tag{1-5}$$

where R and RH are the oxidized and reduced forms of the organic molecule. For such processes the half-wave potential will be a function of pH (with a negative shift of about 59 mV/n for each unit increase in pH, due to decreasing availability of protons). Thus, in organic polarography, good buffering is vital for generating reproducible results.

The limiting diffusion current, i_d, is directly proportional to the concentration of the electroactive species by the Ilkovic equation:

$$i_d = 607 \, nD^{1/2} \, m^{2/3} \, t^{1/6} \, C \tag{1-6}$$

where m and t are the rate of mercury flow and drop time, respectively. To determine the diffusion current, it is necessary to subtract the residual current. This can be achieved by extrapolating the residual current prior to the wave or by recording the response of the deaereated supporting electrolyte solution. Standard addition or a calibration curve is often used for quantitation. Polarograms to be compared for this purpose must be recorded in the same way.

Table 1-1
Functional Groups Reducible at the DME

Class of Compounds	Functional Group	$E_{1/2}$,V*
Azo	—N=N—	−0.4
Carbon-carbon double bond†	—C=C—	−2.3
Carbon-carbon triple bond†	—C≡C—	−2.3
Carbonyl	C—O	−2.2
Disulfide	S—S	−0.3
Nitro	NO$_2$	−0.9
Organic halides	C—X (X = Br,Cl,I)	−1.5
Quinone	C=O	−0.1

*Against the saturated calomel electrode at pH = 7.

†Conjugated with a similar bond or with an aromatic ring.

For reversible systems (with fast electron-transfer kinetics), the shape of the wave may be described by

$$E = E_{1/2} + 0.059/n \ \log(i_d - i/i) \tag{1-7}$$

It follows from Equation 1-7 that a plot of E versus $\log (i_d - i/i)$ should yield a straight line with a slope of $0.059/n$ at 25°C. Such a plot offers a method for the determination of n. In addition, the intercept of this line will be the half-wave potential. It should be emphasized that many polarographic processes, especially those of organic compounds, are not reversible. For those that depart from reversibility, the wave is "drawn out," with the current not rising steeply, as is shown in Figure 1-3.

When the sample solution contains more than one reducible species, diffusion currents resulting from each of them are observed. The heights of the successive waves can be used to measure the individual analytes, provided there is a reasonable difference (>0.2 V) between the half-wave potentials. The baseline for measuring the limiting current of the second species is obtained by extrapolation of the limiting current of the first process. With a potential window of about 2 V, five to seven individual polarographic waves could be observed. Solution parameters, such as the pH or concentration of complexing agents, can be manipulated to improve the resolution of two successive waves. Successive waves are observed also for samples containing a single analyte that undergoes reduction in two or more steps (for example, 1,4-benzodiazepine, tetracycline).

The background (residual) current that flows in the absence of the electroactive species of interest is composed of contributions due to double-layer charging and redox reactions of impurities, as well as of the solvent, electrolyte, or electrode. The latter processes (e.g., hydrogen evolution and mercury oxidation) are those that limit the working potential range. In acidic solutions, the negative background limit shifts by approximately 59 mV/pH unit to more positive potentials with decreasing pH. Within the working potential range, the charging current is the major component of the background. It is the current required to charge the electrode-solution interface (which acts as a capacitor) upon changing the potential or the electrode area. Thus, the charging current is present in all conventional polarographic experiments, regardless of the purity of reagents. At low concentrations of the electroactive analyte, about 5×10^{-6} M, the charging current contribution becomes larger than the analytical faradaic current, and the measurement becomes impossible. The optimum concentration range for conventional polarographic measurements is therefore $10^{-2} - 10^{-4}$ M. As a result, in recent times conventional polarography is rarely advocated for low-concentration determinations; pulse voltammetry and related methods are recommended for this purpose.

1-5 PULSE VOLTAMMETRY

Pulse voltammetry has been developed from the desire to suppress the charging background current and hence lower the detection limits of voltammetric measurements. Several waveforms are employed for improving the faradaic-to-charging-current ratio, including staircase, normal pulse, differential pulse, and square wave. All are based on what is called *chronoamperometry*, the measurement of current as a function of time after applying a potential pulse (step). The duration of the pulse is about 50 ms. After the potential is stepped, the charging current decays very rapidly to a negligible value, while the faradaic current decays more slowly. Thus, by sampling the current in the last few milliseconds of the pulse, the current is almost purely faradaic.

With both normal-pulse and differential-pulse voltammetry, one potential pulse is applied for each drop of mercury when the DME is used (both techniques can also be used at solid electrodes). By controlling the drop time (with a mechanical knocker) the pulse is synchronized with the maximum growth of the mercury drop. At this point, near the end of the drop lifetime, the contribution of the charging current is minimal, while the faradaic current reaches its maximum value (based on the time dependencies of these components).

1-5.1 Normal-Pulse Voltammetry

Normal-pulse voltammetry consists of a series of pulses of increasing amplitude applied to successive drops at a preselected time near the end of each drop lifetime (Figure 1-4). Between the pulses, the electrode is kept at a constant (base) potential at which no reaction of the analyte occurs. The amplitude of the pulse increases linearly with each drop. The current is measured about 40 ms after the pulse is

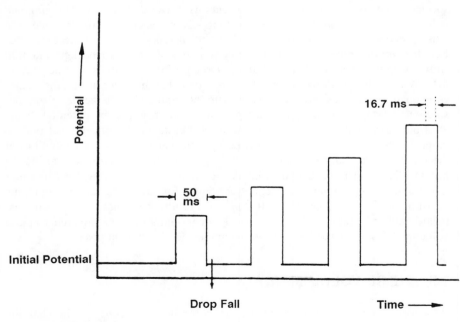

FIGURE 1-4
Excitation signal for normal-pulse voltammetry.

applied, at which time the contribution of the charging current is nearly zero. In addition, because of the short pulse duration, the flux of analyte particles is larger, and hence the faradaic current is increased (five to seven times compared with conventional polarography). This procedure yields a wave-shaped polarogram at concentrations as low as 5×10^{-7} M. More sensitive and useful potential-pulse techniques are differential-pulse and square-wave voltammetry.

1-5.2 Differential-Pulse Voltammetry

Differential-pulse voltammetry has been extremely useful for trace measurements of electroactive species in body fluids and tissues. In differential-pulse voltammetry, fixed-magnitude pulses—superimposed on a linear potential ramp—are applied to the working electrode at a time just before the end of the drop (Figure 1-5). The current is sampled twice, just before the pulse application and just before the end of the pulse. The first current is instrumentally subtracted from the second, and this current difference is plotted versus the applied potential. The resulting differential-pulse voltammogram consists of current peaks, the heights of which are directly proportional to the concentration of the corresponding analytes. The peak potential (E_p) can be used to identify the species as it occurs near the polarographic half-wave potential:

$$E_p = E_{1/2} - \Delta E/2 \tag{1-8}$$

where ΔE is the pulse amplitude. This operation results in effective correction of the charging background current, thus allowing measurements at concentrations as low as 10^{-8} M (about 1 ng/ml sample). The improved detectability over dc polarography is demonstrated in Figure 1-6, which compares the response of both techniques for the antibiotic chloramphenicol present at the $1.3 \times 10^{-5} M$ level.

The peak-shaped response of differential-pulse measurements results also in improved resolution between two species with similar redox potentials. In various situations, peaks separated by 50 mV may be detected. This feature, and the flat background current, makes the technique particularly useful for analysis of mixtures.

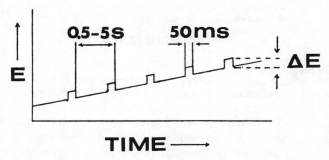

FIGURE 1-5
Excitation signal for differential-pulse voltammetry.

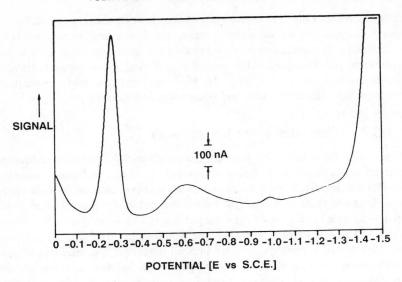

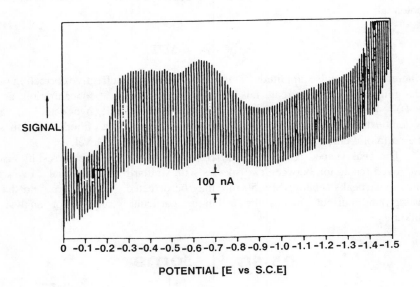

FIGURE 1-6

Differential-pulse (top) and dc (bottom) polarograms for 1.3 $\times 10^{-5}$ M chloramphenicol. (Reproduced with permission.[9])

The selection of the pulse amplitude and potential scan rate usually requires a tradeoff among sensitivity, resolution, and speed. For example, larger pulse amplitudes result in larger and broader peaks. Pulse amplitudes of 25–50 mV coupled with a 5-mV/s scan rate are commonly employed. Irreversible redox systems result in lower and broader current peaks (i.e., inferior sensitivity and resolution) compared with those predicted for reversible systems.[10] In addition to improvements in sensitivity and resolution, the technique can provide information about the chemical form in which the analyte appears (oxidation states, complexation, etc.).

Because of the above advantages and the availability of low-cost instruments, differential-pulse voltammetry is often the method of choice for analysis of body fluids. For example, the inherently low detection limits of the method permit measurements of drugs in body fluids following therapeutic administration on a routine basis. The majority of all papers reporting voltammetric drug analysis present results obtained by the differential-pulse waveform. Numerous applications, based primarily on the dropping mercury and carbon paste electrodes, are described in Section 1-13.

1-5.3 Square-Wave Voltammetry

Square-wave voltammetry is a large-amplitude differential technique in which a waveform composed of a symmetrical square wave, superimposed on a staircase, is applied to the working electrode[11] (Figure 1-7). The current is sampled twice during each square-wave cycle, once at the end of the forward pulse and once at the end of the reverse pulse. The difference between the two measurements is plotted versus the staircase potential. The resulting peak-shaped voltammogram displays excellent sensitivity and effective discrimination against background contributions (detection limits near 1×10^{-8} M). The major advantage of square-wave voltammetry is its speed. Frequencies of 1 to 100 square-wave cycles per second permit the use of extremely fast potential scan rates. As a result, the analysis time is drastically reduced; a complete polarogram can be recorded within a few seconds, as compared with about 3 min in differential-pulse voltammetry. Due to the fast scan rates, the entire voltammogram is recorded on a single mercury drop. The speed of the technique allows increased sample throughput, as desired in many clinical situations. For example, in response to the needs of greater speed and automation, EG&G PAR Inc. has recently introduced a new polarographic system (Model 309) that couples the rapid scanning and sensitivity capabilities of square-wave voltammetry with automatic sample introduction and rapid deaeration (in a nebulization process).

1-6 AC VOLTAMMETRY

In ac voltammetry, a constant sinusoidal ac potential is superimposed upon a dc potential ramp. Usually the alternating potential has a frequency of 50–100 Hz and an amplitude of 10–20 mV. The resulting ac current is displayed versus the potential. Such a voltammogram shows a peak, the potential of which is the same as the polarographic half-wave potential. (At this region the sinusoid has maximum

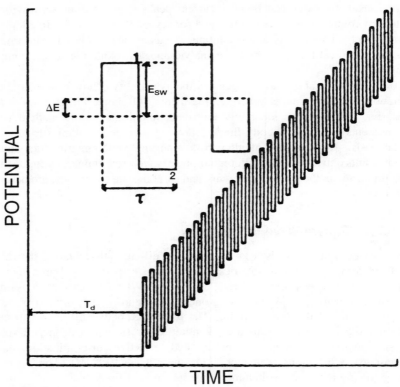

FIGURE 1-7
Square-wave waveform showing the step height, ΔE; amplitude, E_{sw}; square-wave period, τ; delay time, T_d; and current measurement times, 1 and 2. (Reproduced with permission.[12])

impact on the surface concentration, i.e., on the current.) The height of the ac voltammetric peak is proportional to the concentration of the analyte and, for a reversible reaction, to the square root of the frequency.

The detection of the ac component allows one to separate the contributions of the faradaic and charging currents because of the phase difference between them. The charging current is thus rejected using a phase-sensitive lock-in amplifier. As a result, reversible electrode reactions yield detection limits around $5 \times 10^{-7} M$. Substantial loss in sensitivity is expected for analytes with slow electron-transfer kinetics. This may be advantageous for measurements of species with fast electron-transfer kinetics in the presence of one (e.g., dissolved oxygen) that is irreversible. (For the same reason, the technique is very useful for the study of electron processes.) Theoretical discussions on ac voltammetry are available in the literature.[13,14] Practical applications of the technique for measuring pharmaceutically important compounds are numerous.[15] One of the most useful bioanalytical applications of ac polarography is to determine surface-active compounds via their tensammetric response, i.e., two sharp peaks associated with their adsorption and desorption processes.

1-7 CHRONOAMPEROMETRY

Chronoamperometry involves stepping the potential of the working electrode from a value at which no faradaic reaction occurs to a potential at which the surface concentration of the electroactive species is effectively zero (Figure 1-8A). A stationary working electrode and unstirred solution are used. The resulting current-time dependence is monitored. As mass transport under these conditions is solely by diffusion, the current-time curve will exhibit an exponential decrease of current with time (Figure 1-8B). The fundamental equation, which reflects the flux

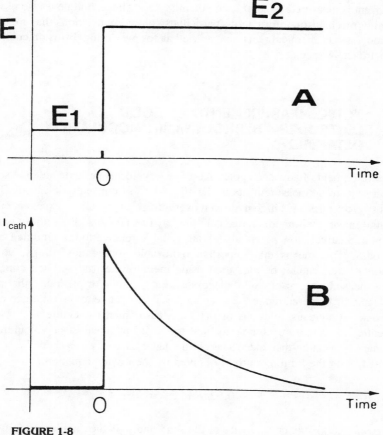

FIGURE 1-8
Chronoamperometry. (A) Potential-time waveform; (B) current-time response.

of material to the surface, is given by

$$i(t) = nFAD^{1/2}C/\pi^{1/2}t^{1/2} \qquad (1-9)$$

This is the Cottrell equation, an important relationship, fundamental to many voltammetric systems with diffusion control. This equation can be squeezed down to its essentials

$$i(t) = KCt^{-1/2} \qquad (1-10)$$

The major bioanalytical significance of chronoamperometry is in in-vivo monitoring of neurotransmitters (Chapter 5). In such studies, the potential of the implantable electrode is repetitively pulsed at fixed time intervals (e.g., 1 min), and the current is measured 1 s after each potential step. The technique is also useful for obtaining mechanistic information (particularly on redox reactions that are coupled to homogeneous chemical reactions) as well as for measuring diffusion coefficients of electroactive species.

1-8   ANODIC MEASUREMENTS AT SOLID ELECTRODES—HYDRODYNAMIC MODULATION VOLTAMMETRY

The development of advanced potential-pulse waveforms has extended the scope of mercury electrode measurements to the $10^{-6}-10^{-8}M$ concentration level. Though mercury electrodes have high hydrogen overvoltage and hence an excellent cathodic potential range, their anodic range is limited by the mercury (dissolution) oxidation current at relatively low potentials (about $+0.4$ V relative to the saturated calomel electrode). This means that important oxidizable compounds (drugs, vitamins, hormones, etc.) cannot be measured using mercury electrodes. Such compounds can be measured at various solid electrodes, e.g., carbon or platinum, that possess a wide anodic potential range (up to about $+1.1$ V, relative to the saturated calomel electrode, in aqueous solutions of pH 7). Measurements over the $10^{-3}-10^{-5}$-M concentration range are commonly performed using linear scan voltammetry at stationary or rotating-disk electrodes. In the latter case, a wave-shaped response is obtained, with the limiting current i_L given by the Levich equation:

$$i_L = 0.62nFACD^{2/3}\upsilon^{-1/6}w^{1/2} \qquad (1-11)$$

where υ is the kinematic viscosity of the fluid and w is the angular velocity of the disk. Besides quantitation, rotating-disk experiments can be employed for mechanistic investigations, particularly when using a rotating-ring–disk configuration.

For trace measurements, more sensitive schemes are used. The coupling of potential-pulse procedures with solid electrodes offers inferior sensitivity compared with that of analogous measurements at mercury electrodes. This is due to additional background current components associated with redox reactions of solid electrode surfaces. Potential-pulse waveforms do not correct against these background contributions, thus yielding relatively high detection limits (10^{-6}–10^{-7} M). Carbon electrodes (particularly carbon paste) are preferred over metallic ones for potential-pulse measurements.

One approach to correct the background current at solid electrodes and thus to lower the detection limits is hydrodynamic modulation voltammetry.[16] The technique is based on pulsing the convection rate near the electrode surface and measuring the corresponding current difference. Hence, a relatively small transport-controlled analytical current can be observed without interference from large contributions of charging, surface, and irreversible solvent decomposition background currents (of a nonconvective nature). The most widely used form of this method is sinusoidal hydrodynamic modulation, at the rotating-disk electrode, which utilizes sinusoidal speed programming around a fixed center value.[17] Other forms have applied pulsed-rotation,[18] pulsed-flow,[19] or stopped-rotation[20] voltammetric schemes. Other sensitive solid electrode procedures, based on controlled adsorptive-extractive accumulation of the analyte followed by voltammetric quantitation of the surface species, are described in the following sections. Most solid electrodes exhibit interesting surface phenomena that can significantly influence the redox behavior of the analyte. Hence, surface treatment procedures can be used to enhance the voltammetric response of interest. The monograph *Electrochemistry at Solid Electrodes*[1] provides an excellent discussion of the use of solid electrodes. Besides their utility for conventional voltammetric measurements, solid electrodes are extremely useful for amperometric detection following liquid chromatography and in-vivo monitoring. These subjects are reviewed in Chapters 4 and 5 of this volume.

1-9 STRIPPING ANALYSIS

Stripping analysis is an extremely sensitive electrochemical technique for measuring trace metals. Its remarkable sensitivity is attributed to the combination of an effective preconcentration step with advanced measurement procedures that generates an extremely favorable signal-to-background ratio. As a result, four to six metals can be measured simultaneously in biological matrices at concentration levels down to 10^{-10} M, utilizing relatively inexpensive instrumentation.

Essentially, stripping analysis is a two-step technique. The first, or deposition step, involves the electrolytic deposition of a small portion of the metal ions in solution into the mercury electrode to preconcentrate the metals. This is followed by the stripping step (the measurement step), which involves the dissolution (stripping) of the deposit. Because of their inherent sensitivity and convenience, the rotating thin mercury film (on glassy carbon) electrode and the hanging mercury drop electrode are the most practical electrodes for stripping analysis.

Anodic stripping voltammetry is the most widely used form of stripping analysis. In this case, the preconcentration is done by cathodic deposition at a controlled time and potential [more negative than the reduction potential of the element(s) to be measured]. The metal ions reach the electrode surface by diffusion and convection—forced by rotating the working electrode or stirring the solution—where they are reduced and concentrated as amalgams in the small-volume mercury electrode:

$$M^{n+} + Hg + ne^- \rightleftharpoons M(Hg) \tag{1-12}$$

The duration of the deposition step is selected according to the concentration level of the metal ions in question: from less than 1 min at the $10^{-7}M$ level to about 20 min at the $10^{-9}M$ level.

Following the preselected time of the deposition step, the potential is scanned anodically, linearly, or in a more sensitive potential-time waveform (usually the differential-pulse ramp). During this scan, the amalgamated metals are stripped out of the electrode in an order that is a function of each metal standard potential, are reoxidized, and give rise to anodic peak currents that are measured. The potential-time sequence used in such experiments, along with the resulting response, is shown in Figure 1-9. The resulting peak current, which is proportional to the concentration of the metal ion in the bulk solution, depends upon various parameters of the deposition and stripping steps (deposition time, mass transport, scan rate), as well as on the characteristics of the metal ion (diffusion coefficient, number of electrons) and the electrode geometry.

Other versions of stripping analysis include potentiometric stripping and cathodic stripping voltammetry. Potentiometric stripping analysis differs from anodic stripping voltammetry in the method used for reoxidation of the amalgamated metals. In this case, the potentiostatic control is disconnected following the preconcentration, and the concentrated metals are reoxidized by an oxidizing agent, for example, O_2, Hg(II), which is present in the solution, or by applying an appropriate constant current. During the oxidation process, the variation of the working electrode potential is recorded as a function of time. The resulting potential versus time response consists of stripping plateaus, as in a redox titration curve; the time lapse between two consecutive equivalent points is taken as a measure of the sample concentration of the particular analyte oxidized in this interval. Details of the theoretical foundation of potentiometric stripping analysis are given in Reference 21.

In cathodic stripping voltammetry, the oxidation of the analyte is used for its preconcentration as an insoluble film on the electrode; subsequently, the concentrated analyte is being reduced and measured during a negative-going scan:

$$A^{n-} + Hg \rightleftharpoons HgA + ne^- \tag{1-13}$$

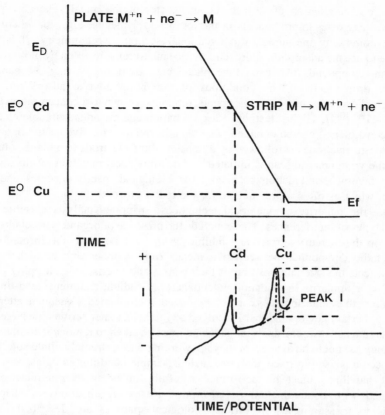

FIGURE 1-9
The potential-time sequence used in a linear scan anodic stripping experiment along with the resulting voltammogram.

(This mode is thus the "mirror image" of anodic stripping voltammetry.) A^{n-} is a univalent (e.g., Cl^-) or divalent (e.g., S^{2-}) anion; in the former case HgA corresponds to the mercurous salt, $Hg_2 Cl_2$. Various organic and inorganic analytes, e.g., thiols or halide ions, capable of forming insoluble salts with mercury have been measured by cathodic stripping voltammetry.

Stripping analysis is useful when the metal ion is readily reduced to the metallic state and reoxidized. About 20 amalgam-forming metals, including lead, cadmium, zinc, copper, bismuth, thallium, indium, antimony, tin, nickel, manganese, and gallium, are easily measurable by stripping analysis with mercury electrodes. In addition, metal ions such as selenium, arsenic, mercury, and gold are determined at bare solid electrodes such as carbon or gold. Trace levels of additional metals, such as aluminum, iron, vanadium, cobalt, molybdenum, uranium, chromium, titanium, and lanthanum, can be monitored using the recently introduced adsorptive stripping voltammetry.[22] Adsorptive stripping voltammetry involves the formation and adsorptive accumulation of surface-active chelates of the metal. The adsorptive approach may offer improvements in the response for metals, e.g., nickel, tin, or copper, commonly measured by conventional stripping voltammetry. It is also possible to use the adsorption approach for trace measurements of many biologically important compounds that cannot be plated electrolytically. Using the hanging mercury drop electrode and conditions of maximum analyte adsorption, this approach allows measurements of organic compounds down to a concentration level of 10^{-9}–10^{-10} M. High selectivity (discrimination against nonadsorbable electroactive constituents) can be obtained when an adsorptive-extractive accumulation at carbon paste electrodes is followed by a transfer of the electrode to a blank solution where the voltammetric scan is initiated.[23] Adsorptive accumulation results also in a more favorable interaction between large biological macromolecules and the surface, which is often a prerequisite for the redox process.

Like the majority of analytical techniques, stripping analysis is subject to various types of interferences. These include the presence of organic surfactants that adsorb on the mercury surface and inhibit the metal deposition, the formation of intermetallic compounds between two metals (e.g., copper-zinc or nickel-zinc) which affects the size and position of their peaks, and overlapping stripping peaks caused by a similarity in oxidation potentials (e.g., indium-cadmium, lead-tin). In most cases, these interferences can be minimized or eliminated if adequate attention is paid to certain key operations. Dissolved oxygen is another serious interferent in most versions of stripping analysis, and care must be taken to remove it completely.

Analysts should be aware that they are working with extremely dilute solutions. All procedures used in trace analysis, such as sample handling and glassware and reagent handling, must be performed carefully in order to minimize errors associated with contamination or adsorption losses. Applications of stripping analysis to measurements in various biological matrices are described in the following sections. For a more detailed description of the technique the reader is referred to a recent monograph.[24]

1-10 CYCLIC VOLTAMMETRY

Cyclic voltammetry is the most widely used technique for acquiring qualitative information about electrochemical reactions. It is a powerful tool for the rapid determination of formal potentials, detection of chemical reactions that precede or follow electron transfer, or evaluation of electron-transfer kinetics. Accordingly, the technique has been widely used in studying the redox mechanisms of many biologically significant molecules. The results of such investigations into the redox chemistry of biomolecules and drugs might have profound effects on our understanding of their in-vivo redox processes or pharmaceutical activity. Cyclic voltammetry is often the first experiment performed in an electroanalytical study.

Cyclic voltammetry consists of cycling the potential of the working electrode — which is immersed in an unstirred solution — over a given range and measuring the resulting current. Depending on the information sought, single or multiple cycles can be used.

Figure 1-10 illustrates the expected response of a reversible redox couple during a single potential cycle. It is assumed that only the reduced form R is present initially. Thus, a positive-going potential scan is chosen for the first half-cycle, during which an anodic current peak is observed. Because the solution is quiescent, the product O generated during the forward scan is available at the surface for reduction during the reverse scan, resulting in a cathodic current peak. The important parameters of the resulting cyclic voltammogram are the magnitudes of the anodic peak current, $i_{p,a}$; the cathodic peak current, $i_{p,c}$; and the positions of the peaks along the potential axis, $E_{p,a}$ and $E_{p,c}$. For a simple reversible system with no chemical complications, the ratio of the peak currents, $i_{p,c} / i_{p,a}$, is unity, with the peak current given by

$$i_p = 269 \, n^{3/2} \, A D^{1/2} \, Cv \tag{1-14}$$

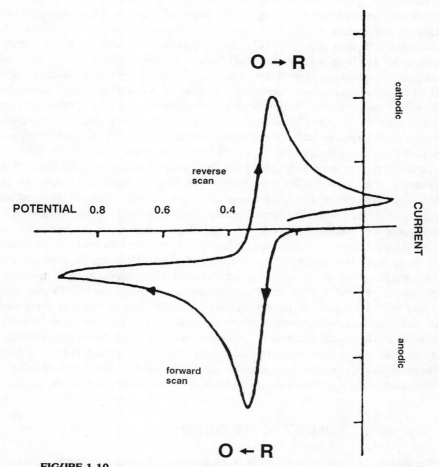

FIGURE 1-10
Typical cyclic voltammogram for a reversible $O + ne^- \rightleftharpoons R$ redox process.

where A is the electrode area and v is the scan rate. (As discussed below, the ratio of peak currents can be significantly influenced by chemical reactions coupled to the redox process.) The potentials of the two peaks are separated by a small increment

$$\Delta E_p = E_{p,a} - E_{p,c} = 0.059/n \qquad (1\text{-}15)$$

This potential difference can be used to determine the number of electrons transferred in the redox reaction.

The situation is very different when the redox reaction is not reversible, when chemical reactions are coupled to the redox process, of when adsorption of the reactant and/or product occurs. Actually, it is these "nonideal" processes that are usually of greatest chemical interest and for which the diagnostic power of cyclic voltammetry is very useful. For irreversible processes (those with sluggish electron-transfer kinetics), the individual peaks are reduced in size and widely separated. A great deal of diagnostic information regarding these "nonideal" processes can be obtained by changing the potential scan rate. This is especially the case for redox reactions that are coupled to homogeneous chemical reactions. The occurrence of such preceding or succeeding chemical reactions, which directly affect the available concentration of the electroactive species at the electrode surface, is common to many redox processes of biologically important compounds. In such situations, cyclic voltammetry is able to generate a species during the forward scan, and probe its fate within a few seconds from the reverse scan and subsequent cycles. For example, the tranquilizer chlorpromazine undergoes oxidation to form a radical cation that reacts with water to form sulfoxide. Because sulfoxide is not electroactive, a smaller reduction peak is observed. The rate constant for such coupled chemical reactions can be estimated from the effect of the potential scan rate. Cyclic voltammetry can also be used for evaluating the interfacial behavior of biological compounds. For example, Figure 1-11 illustrates repetitive cyclic voltammograms, at the hanging mercury drop electrode, for riboflavin in a sodium hydroxide solution. A gradual increase of the anodic and cathodic peak currents is observed, indicating progressive adsorptive accumulation at the surface. The separation between the peak potentials is smaller than expected for solution-phase processes. The quantity of charge consumed by the surface process at saturation (peak area) can be used to calculate the surface coverage. These and other interfacial studies may provide better understanding of interactions of biological compounds with biosurfaces (and thus their biological function). Unlike conventional electrodes, ultramicroelectrodes exhibit sigmoidal-shaped cyclic voltammograms and permit use of ultrafast scan rates that can be exploited for rapid kinetics studies. For more detailed information on interpretation of cyclic voltammograms and discussion of the technique, see References 4, 26, and 27.

1-11 SPECTROELECTROCHEMISTRY

The coupling of optical and electrochemical methods, spectroelectrochemistry, has been employed for several years to investigate a wide variety of biological redox systems. This methodology combines electrochemical perturbation by an optically

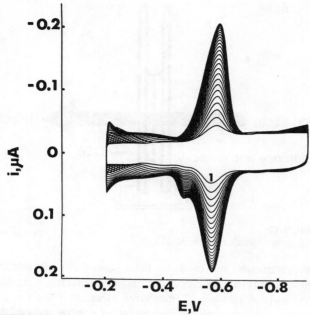

FIGURE 1-11
Repetitive cyclic voltammograms for 1 μM riboflavin in a 0.001 M NaOH solution. (Reproduced with permission.[25])

transparent electrode (OTE) with simultaneous optical monitoring of the solution adjacent to the electrode surface. The types of knowledge that can be obtained in such experiments include (a) redox potentials of electroactive species, (b) the number of electrons involved in the redox reaction, (c) the spectra of electrogenerated species, and (d) optical monitoring of preceding or following chemical reactions.

A variety of optical methods have been coupled with electrochemistry. The most commonly used is ultraviolet-visible absorption spectroscopy. In a typical experiment, one measures absorbance changes resulting from species produced or consumed in the redox process. A simple optically transparent thin-layer cell used in such experiments includes gold minigrid or porous reticulated vitreous carbon transparent materials that serve as the working electrode. This is sandwiched between two optical-quality quartz microscopic slides (Figure 1-12). The resulting chamber, containing the electroactive species in solution, contacts a larger container that holds the reference and auxiliary electrodes. The optical beam of the spectrophotometer is passed directly through the transparent electrode and the solution. The working volume of the cell is only 30–50 μl, and complete electrolysis of the solute requires only 30–60 s. These features of transparent thin-layer cells have been extremely useful for studying the optical characteristics of electrogenerated species. For measurements of formal redox potentials ($E°$) and electron-transfer stoichiometry (n values), the redox couple is converted incrementally from one oxidation state to another by a series of applied potentials.

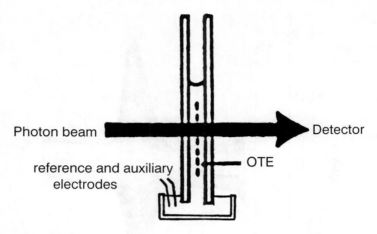

FIGURE 1-12
Thin-layer spectroelectrochemical cell.

Absorption measurements are used to determine the concentration ratio of oxidized-to-reduced ([O]/[R]) species at each applied potential. One can obtain the formal potential and the number of transferred electrons from the intercept and slope, respectively, of the resulting linear plot of applied potential versus log ([O]/[R]). For many biological macromolecules that undergo electron transfer very slowly at the electrode, one can add a smaller electroactive species (known as a "mediator") to transfer the electrons between the biocomponent and the electrode.

Besides these transmission procedures, powerful spectroelectrochemical data can be obtained in experiments in which the light beam is reflected from the electrode surface, i.e., techniques of ellipsometry or specular reflectance.

Good examples of the utility of spectroelectrochemistry for studying biological systems include estimating $E°$ and n of blue copper proteins,[28] studying the heterogeneous electron transfer of cytochrome C,[29] monitoring the electrogenerated intermediate in the oxidation of 5,6-diaminouracil,[30] measuring the lifetime of reduced chlorophyll,[31] and studying the charge transfer between a chlorpromazine cation radical and dopamine.[32] Several reviews[33, 34] discuss more completely the principles and applications of spectroelectrochemistry.

1-12 COULOMETRY

The term *coulometry* implies an electroanalytical technique in which the electrical charge Q, needed to convert the analyte quantitatively to a different oxidation state, is measured. This permits a simple calculation of the amount of analyte present, W, based on Faraday's law:

$$W = QM/nF \qquad (1\text{-}16)$$

where M is the formula weight of the analyte. Being an "absolute" method, no chemical standards or calibration curves are required.

Two coulometric procedures are possible: operation at constant potential, during which the current decreases exponentially to practically zero (with the charge being measured by integration of the current-versus-time curve), and operation at constant current, so that the amount of material reacted is proportional to the elapsed time. The second approach is useful for coulometric titrimetry, in which the conventional volumetric addition of a reagent is replaced by electrochemical generation (e.g., electrogeneration of iodine or bromine for assays of ascorbic acid[35] and sulfa drugs,[36] respectively). This procedure results in improved accuracy compared with standard titrimetric methods because small quantities of the titrant can be accurately generated. In addition, titrants that are unstable or troublesome in volumetric methods, e.g., halogens, can be electrochemically generated. The utility of coulometric titrimetry for determination of pharmaceutical compounds was reviewed by Patriarche.[37]

The controlled-potential coulometric approach is useful in the determination of small amounts of electroactive species where separation by potential is advantageous.[38] This procedure has been used to elucidate the nature of electrode reactions (e.g., n values of the reduction of chlordiazepoxide,[39] by electrolyzing a known amount of the compound), to determine biological compounds in various matrices (e.g., vitamin K in pharmaceutical tablets and biological fluids[40]), and to evaluate various standard materials.

The choice of electrolysis conditions defines the selectivity of coulometric assays. Some factors that must be considered include instrumentation, cell geometry, electrodes, solvent and electrolyte, analyte concentration, and potential or current (depending upon the procedure used). Some of these conditions can be selected from preliminary voltammetric or other electrochemical experiments.

Table 1-2 summarizes the characteristics of modern voltammetric methods,

Table 1-2
Properties of Voltammetric Techniques

Technique*	Working Electrode†	Detection Limit, M	Speed (time per cycle), min	Response Shape
DC polarography	DME	10^{-5}	3	Wave
NP polarography	DME	5×10^{-7}	3	Wave
DP polarography	DME	10^{-8}	3	Peak
DP voltammetry	Solid	5×10^{-7}	3	Peak
SW polarography	DME	10^{-8}	0.1	Peak
AC polarography	DME	5×10^{-7}	1	Peak
Stripping voltammetry	HMDE, MFE	10^{-10}	3–6	Peak
Adsorptive stripping voltammetry	HMDE	10^{-10}	2–5	Peak
Adsorptive stripping voltammetry	Solid	10^{-9}	4–5	Peak

*DC= direct current; NP= normal pulse; DP= differential pulse; SW= square wave; AC = alternating current.

†HMDE= hanging mercury drop electrode; MFE= mercury film electrode.

including comparison of their sensitivity. The following sections discuss applications of these techniques to problems of clinical and biomedical analysis.

1-13 CLINICAL APPLICATIONS OF VOLTAMMETRIC TECHNIQUES

Having examined the basic voltammetric techniques, we now review their potential for problems of clinical interest. There is no real shortage of applications of voltammetric techniques to clinical medicine. An extremely large number of species of clinical interest are either reducible at mercury electrodes or oxidizable at various solid electrodes. These can be divided into endogenous and exogenous organic compounds and toxic and essential elements. While voltammetry of inorganic species typically involves a change in the oxidation state of a central atom (usually a metal), redox reactions of organic compounds include changes in covalent bonds leading to interconversion of functional groups or addition, substitution, elimination, coupling, or cleavage reactions. Therefore, molecular structure is the primary determinant for the voltammetric activity of an organic compound. This can be deduced from an examination of the functional groups present. The functional groups that exhibit excellent voltammetric behavior include quinone, nitro, nitroso, azo, carbonyl, imine, activated double bonds, sulfhydryl, halogen, amine, phenol, azine, and amide. Many compounds of biological interest contain more than one electroactive site, e.g., anthracycline antibiotics with a reducible quinone group and an oxidizable hydroquinone center. Bioelectroanalytical methods for the determination of these compounds can be based on either their oxidation or their reduction. A detailed description of mechanisms of redox reactions involving these functionalities is beyond the scope of this discussion. Interested readers are referred to the literature on organic electrochemistry.[41] The redox activity depends not only on the functional group involved but also on its environment and the molecular skeleton. If a compound undergoes more than one redox process, the position of successive peaks (or waves), as well as their relative magnitudes, can be very diagnostic.

In contrast to inorganic species, the reactions of organic compounds are usually irreversible and often proceed in several steps. Because of the involvement of protons in most organic redox reactions (Equation 1-5), the position of the wave (or peak) along the potential axis, as well as the reaction mechanism, is affected by the pH. Measurement of nonelectroactive compounds may be feasible via an appropriate derivatization, i.e., the introduction of electroactive functionality by a chemical reaction preceding the voltammetric analysis.

Whereas the appeal of voltammetric techniques to clinical medicine is attributed to their sensitivity and simplicity, improved specificity and stability are frequently obtained via extraction or cleanup procedures (which are routine in bioanalytical laboratories). In some cases, direct voltammetric measurements have been performed in untreated biological samples. The coupling of voltammetric techniques with highly specific procedures, such as liquid chromatography or immunoassay, results in powerful analytical tools, as will be described later in this book. Also to be discussed in a separate chapter is the utility of voltammetric techniques for in-vivo measurements.

This section demonstrates the suitability of voltammetric techniques to a wide variety of clinical and biomedical problems. Applications are classified according to physiological activity rather than according to techniques applied.

1-13.1 Trace Elements

The need for monitoring toxic and essential elements in biological samples has led to an increasing demand for suitable analytical techniques. Techniques used for this purpose include neutron activation, atomic absorption, plasma emission or x-ray fluorescence spectroscopy, and voltammetry. Voltammetric techniques, particularly stripping analysis and differential-pulse voltammetry, are very suitable for surveillance of trace metals in humans as these techniques offer the advantages of very low detection limits, multielement capability, and low cost. Of toxicological interest is the ability of voltammetric techniques to differentiate between oxidation states of a given metal and to measure the labile (reactive) form of the metal. Important speciation information is obtained also from the observed shifts in the metal peak (or wave) potential in the presence of various complexing agents.

For more than 15 years, stripping analysis has measured trace metals in body fluids and tissues. Pulse polarographic techniques have been used for metals present at concentrations higher than $1 \times 10^{-7} M$ or for those not measurable by stripping analysis. The voltammetric assay is normally preceded by a pretreatment step that releases the trace metals bound to the sample components and minimizes problems of background current or adsorption of matrix constituents on the working electrode. The two types of decomposition procedures often used for this purpose are wet digestion (with acid mixtures such as nitric and sulfuric acids) and dry ashing. The suitability of these pretreatment procedures for voltammetric analysis of biological materials was evaluated in various studies.[42-45] Simpler (and faster) procedures based on the use of an ion-exchange reagent, Metexchange,[46] or adsorptive removal of parts of the organic matrix on an apolar chromatographic medium[47] have been found useful in various clinical situations. Simon et al.[48] have compared methods for elimination of organic matter from urine. As in all trace techniques, judicious precautions must be taken to minimize risks of contamination or loss of the analyte. For example, reagents used for both decomposition and determination should be of the highest purity possible. Because of the considerable amounts of salts present in biological materials, addition of an electrolyte may be avoided with dry ashed samples.[48] Problems encountered and precautions to be taken in trace element analysis of biological samples have been reviewed recently.[49]

Schemes for multielement determination in biological samples, based on a combination of stripping voltammetry and pulse polarographic techniques, have been suggested by Nürnberg,[50] Bond,[51] and others.[52] For example, simultaneous determination of selenium, lead, cadmium, copper, zinc, cobalt, and nickel, present in a single sample at different concentrations, can be accomplished by employing techniques such as anodic or cathodic stripping voltammetry and differential-pulse polarography, together with simple chemical manipulations (Figure 1-13). The use of microprocessor-controlled instrumentation and the static mercury drop electrode enables convenient change from one technique to another. Quantification is usually

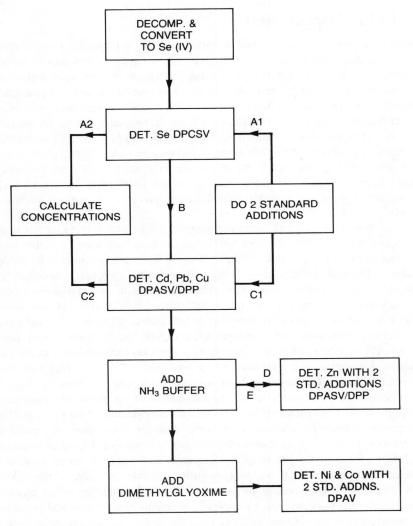

FIGURE 1-13

Analytical scheme for the sequential simultaneous determination of selenium, cadmium, lead, copper, zinc, nickel, and cobalt in a single sample solution by voltammetric analysis. (Reproduced with permission.[51])

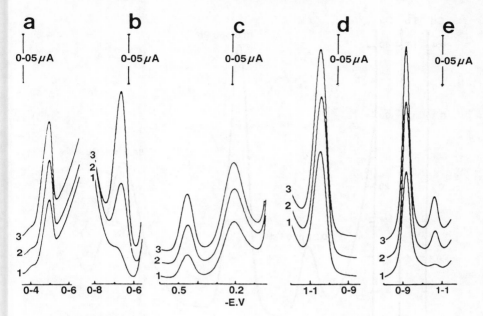

FIGURE 1-14

Sequentual simultaneous determination of selenium (a), cadmium (b), lead and copper (c), zinc (d), and nickel and cobalt (e) in human urine. The stripping modes used were cathodic stripping (a), anodic stripping (b–d), and adsorptive stripping (e). Curves 2 and 3 represent the response following standard additions. (Reproduced with permission.[51])

accomplished via two standard additions. The utility of such a scheme for the determination of seven metals in human urine sample is demonstrated in Figure 1-14. Additional metals, such as arsenic, thallium, mercury, bismuth, or gold, have been measured in biological samples using different voltammetric procedures (as the following describes).

Blood and urine samples have been widely analyzed for their trace metal content by stripping analysis and pulse polarographic techniques. Various stripping modes have been used for quantification at the trace and ultratrace levels. These include differential-pulse,[50,53] staircase,[46,54] and ac[55] waveforms, as well as potentiometric stripping analysis.[56] Figure 1-15 gives an example of the response obtained for simultaneous determination of cadmium, lead, and copper in human whole blood by differential-pulse anodic stripping voltammetry. In most cases, the conventional stripping electrodes, hanging mercury drop or thin mercury film, are used along with conventional stripping cells. Heineman and coworkers[6] advocated the use of a thin-layer cell with a mercury-coated graphite electrode and a sample volume of 60 μl for trace metal analysis in blood. Micromercury electrodes can also be used for microliter measurements.[57] Coverage of mercury films with a permselective coating offers good protection against organic surfactants.

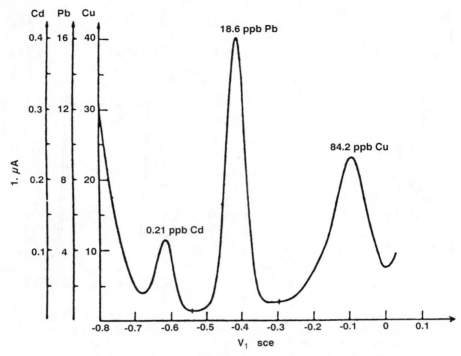

FIGURE 1-15
Simultaneous determination of Cd, Pb, and Cu in an ashed sample
of whole blood by stripping voltammetry at the hanging mercury
drop electrode. Deposition for 10 min at −0.80 V. (Reproduced
with permission.[53])

In addition to assays of blood and urine, stripping analysis enables the determination of trace metals in other biological matrices that indicate exposure to such
metals. These applications include determinations of cadmium, lead, copper, and
zinc in human hair,[58] teeth,[59] eye,[60] or bone.[61] The high accuracy and reliability of such procedures have been confirmed by interlaboratory comparison with
other techniques or by the analysis of NBS standard biological materials. For
example, an interlaboratory comparison of various techniques used for blood lead
analysis, conducted by the Center for Disease Control, ranked stripping voltammetry first, based on overall accuracy and precision.[62] Table 1-3 summarizes the
applications of stripping analysis for the determination of various trace metals
in biological samples. Additional information of such applications is available
in several reviews.[24,53,63] Current research aimed at extending the scope of stripping analysis toward additional metals (based on new adsorptive stripping
procedures[22]) should result in increased use of voltammetry for routine applications
in analytical toxicology. Future work will undoubtedly emphasize the potential of
voltammetry to provide important information regarding the speciation of metals in
body fluids.

Table 1-3
Stripping Analysis of Trace Elements in Body Fluids and Tissues

Analyte Measured and Matrix	Sample Treatment	Stripping Mode	Working Electrode*	Reference
Pb in blood	Wet digestion	Differential pulse	HMDE	64
Cd, Pb in urine	Wet digestion	Differential pulse	MFE	65, 66
Pb in blood	Wet digestion	Differential pulse	MFE, thin-layer cell	6
Cu, Cd, Pb in blood	Ashing	Differential pulse	HMDE	67
Cu, Cd, Pb in urine	None	Differential pulse	HMDE	42
Cu, Pb, Cd in blood	Wet and pressure digestions	Differential pulse	HMDE	68
Cd, Pb in blood and urine	Dilution	Differential pulse	MFE	69
Cd, Pb, Cu, in blood	Dilution	Potentiometric	MFE	56
Pb in blood	Wet digestion	Staircase	MFE	54
As in urine and blood		Staircase	MFE	70
Zn, Cd, Pb, Cu in hair	Wet digestion	Differential pulse	HMDE	58, 71
Zn in eye tissue	Wet digestion		HMDE	60
Cd, Pb, Zn, Cu in teeth	Wet digestion	Differential pulse	HMDE	59, 72
Tl in urine	None	Differential pulse	HMDE, MFE	73
Sb in urine and blood	Wet digestion	Staircase	MFE	74
Co, Ni in biological samples	Dry ashing and chelate formation	Differential pulse	HMDE	75
Ni in nails	Wet digestion and chelate formation	Differential pulse	HMDE	76
Tl in urine	Acidification	AC voltammetry	HMDE	55
Au in blood		Chronopotentiometric	Carbon paste	77
Pb in blood	Ion-exchange reagent	Staircase	MFE	46
Na, K in blood	Solvent optimization	Potentiometric	MFE	78
Hg in urine	Acidification	Potentiometric	Gold	79
Bi in blood and urine	Dilution and ion exchange	Differential pulse	HMDE	80
Pb in urine		Potentiometric	MFE	81
Pb in urine	Acidification	Linear scan	HMDE	82
Zn, Cd, Cu, Pb in hair and nails	Wet digestion	Staircase	MFE	83

Table 1-3 (Continued)
Stripping Analysis of Trace Elements in Body Fluids and Tissues

Analyte Measured and Matrix	Sample Treatment	Stripping Mode	Working Electrode*	Reference
Cu, Pb in urine	Dilution	Coulostatic	MFE	84
Se in biological samples		Cathodic	HMDE	85
Pb, Zn, Cu, Cd in feces	Wet digestion	Differential pulse	HMDE	71
Pb, Cd in bone	Wet digestion	Derivative pulse	HMDE	86

*HMDE = hanging mercury drop electrode; MFE = mercury film electrode.

1-13.2   Measurements of Drugs in Body Fluids

One of the most challenging bioanalytical problems is the determination of drugs and their metabolites in body fluids following the administration of low doses of drugs. The new generation of highly efficacious drugs, which are physiologically active at very low concentrations (10^{-6}–$10^{-9}M$), requires highly sensitive and accurate analytical procedures. The particular method selected for the analysis is based on the sensitivity and specificity requirements of the drug and the characteristics of the assay itself. Voltammetric techniques, particularly differential-pulse voltammetry, are becoming increasingly important for therapeutic monitoring of drugs and metabolites in biological fluids. These techniques are sensitive, reliable, and simple, and the redox reactions often provide selectivity for the compound sought in the presence of degradation products or metabolites. In various situations, the structural information that voltammetric methods have provided in relation to "unknown" metabolites or degradation products has been complementary to the findings of spectroscopic or chromatographic studies.

Separation procedures are frequently necessary prerequisites for the voltammetric measurement of drugs in complex biological matrices. These minimize interferences due to proteins or electroactive compounds and separate of the parent drug from metabolites with similar electroactivity. The most common separation technique is solvent extraction. The pH of the solution to be extracted and the polarity of the organic solvent may have a remarkable effect on the resulting selectivity. Another common pretreatment step is protein precipitation using inorganic salts or organic solvents. The analyte in the solvent extract or protein-free filtrate (PFF) can be determined directly, or following "cleanup," back-extraction, or preconcentration steps. The alternatives for sample preparation that can be used prior to the voltammetric assay are summarized in Figure 1-16. In a few situations, the metabolism reaction occurs close to the electroactive site of the drug, and the voltammetric assay may allow simultaneous determination of the parent compound and the metabolic product.

There are numerous papers dealing with applications of voltammetric techniques to the determination of pharmaceutical compounds in body fluids. Our intention is to select several groups of important drugs to illustrate the clinical utility

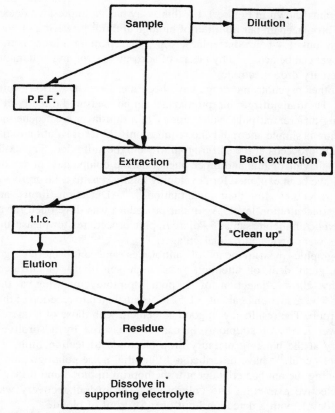

FIGURE 1-16
Sample preparation for voltammetric analysis. (Reproduced with permission.[87])

of voltammetry. For a more complete bibliography the reader is referred to several sources.[4, 87–91]

Polarographic techniques have been used extensively for assays of 1,4-benzodiazepine drugs in physiological fluids, mainly due to the facile reduction of the azomethine group common to these compounds.[87] Such procedures have been useful for the determination of excretion of urinary metabolites of the 1,4-benzodiazepines. For example, bromazepam and its major metabolites can be determined by differential-pulse polarography following selective extraction, with detection limits on the order of 50 ng/ml of urine analyzed.[92] Similarly, chlordeazepoxide and its two major metabolites can be determined in plasma by differential-pulse polarography.[93] The assay involves the selective extraction of the compounds into diethyl ether, followed by thin-layer chromatographic separation of the three compounds. The determination of bromazepam in plasma involves extraction into benzene–methylene chloride; the residue of this extract is dissolved in phosphate buffer (pH 7) prior to the differential-pulse scan.[5] As indicated from the polarograms shown in Figure 1-17, this assay has an impressive detection limit of 10 ng of bromazepam per milliliter of biological fluid. Lowering of the detection limit for various 1,4-benzodiazepines, e.g., diazepam or nitrazepam, to the 1 $\times 10^{-9}M$ level can be achieved by means of adsorptive stripping voltammetry at the hanging mercury drop electrode.[94]

Many other psychotropic drugs have been measured by various voltammetric techniques. The tranquilizer chlorpromazine can be measured directly in urine and plasma using differential-pulse voltammetry in a thin-layer electrochemical cell.[95] The procedure is simple and rapid and requires only 23 μl of solution to fill the cell. The oxidation proceeds at the sulfur atom to yield the sulfoxide. The oxidation and interfacial behavior of this and other phenothiazine compounds at various carbon electrodes have been exploited for developing highly sensitive adsorptive-extractive stripping procedures for their quantitation.[96,97] Detection limits are at the nanomolar concentration level. A similar procedure was developed for measuring several tricyclic antidepressants.[98] Polarographic procedures for measuring psychotropic drugs were reviewed by Oelschlager.[99]

Polarographic measurements of antibiotics and antibacterial agents have received a great deal of attention. The high sensitivity of differential-pulse polarography allows detection of 1 ppm streptomycin using a $0.1M$ NaO solution.[100] The reduction peak, at -1.52 V, is attributed to reaction of the aliphatic aldehyde group. The results are in good agreement with those of a microbiological assay. Lower levels of streptomycin can be measured by adsorptive stripping voltammetry at the hanging mercury drop electrode (detection limit, 7×10^{-10} M).[101] Brooks et al.[102] have described a differential-pulse polarographic procedure for determining the antibacterial agent trimethoprim in blood and urine. The assay involves selective extraction into chloroform. The overall recovery was $81.7 \pm 6.3\%$ (std. dev.), with a detection limit of 0.5 μg/ml of blood.

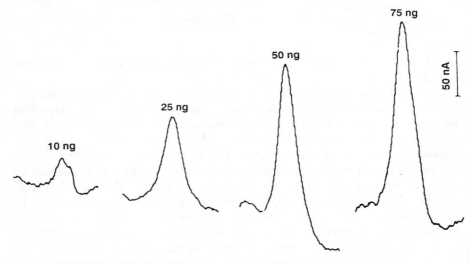

FIGURE 1-17
Differential-pulse polarograms for bromazepam using a 0.5-ml microcell. (Reproduced with permission.[5])

Forsman[103] developed a highly sensitive cathodic stripping procedure for measuring trace amounts of several penicillins. After conversion to the corresponding (electroactive) penicilloic acid, preconcentration proceeds at -0.10 V in the presence of excess copper(II). With a 10-min preconcentration, $2 \times 10^{-10} M$ of penicillin may be detected. The closely related cephalosporin antibiotic undergoes reduction at the dropping mercury electrode without the need for prior functionalization.[104] Well-defined differential-pulse polarographic peaks are obtained at the micromolar concentration level using acidic solutions.

The antibiotic chloramphenicol is particularly suitable for polarographic measurements,[100] owing to the four-electron reduction process of the nitro group. Most drugs and natural compounds do not interfere because of the low potential region at which the reduction occurs. Differential-pulse polarography yields a well-defined peak at the micromolar concentration level (e.g., Figure 1-6). AC and pulse polarography can be used to measure tetracycline antibiotics at the micromolar concentration level.[100] Two peaks, attributed to the reduction of one of the carbonyl groups and a double bond, are observed. An adsorptive stripping procedure, based on controlled interfacial accumulation of tetracycline compounds at the hanging mercury drop electrode, permits convenient measurement at the nanomolar concentration level (e.g., Figure 1-18).[105] The use of polarography for measuring other antibiotics and antibacterial agents has been reviewed by Siegerman.[100]

Voltammetric techniques have been useful for measuring various anticancer and cardiovascular drugs. Differential-pulse polarography can be used for determining the antitumor agent *cis*-dichlorodiammineplatinum(II) in urine.[106] The method is based on isolation of the platinum as its complex with sodium diethyldithiocarbamate and yields a detection limit of 10 ng/ml. Urinary platinum levels can also be determined by differential-pulse voltammetry after deposition onto a rotating glassy carbon electrode.[107] Another widely used antitumor agent, adriamycin, can be measured by adsorptive stripping voltammetry at carbon paste electrodes.[108] No sample treatment is required for analysis of urine samples from cancer patients actively undergoing therapy. The reduction of the adriamycin quinone groups has been used for direct differential-pulse polarographic measurements of the drug in untreated plasma (Figure 1-19).[109] The indole alkaloids, vinblastine and vindoline, can be measured by differential-pulse voltammetry at carbon paste and hanging mercury drop electrodes, down to the micromolar and nanomolar concentration levels, respectively.[110, 111] The reductive procedure allows reliable measurements in body fluids, such as plasma and urine. A differential-pulse polarographic method for the determination of mitomycin C in plasma and urine has been presented.[112] The procedure yields a detection limit of 25 ng/ml, and is applicable to pharmacokinetic studies. A typical polarogram for a spiked plasma is shown in Figure 1-19. Sensitive voltammetric methods for methotrexate and mitomycin C, based on controlled adsorptive accumulation at the hanging mercury drop electrode, offer a detection limit of $2 \times 10^{-9} M$.[113, 114] The cardiac glycosides digoxin and digitoxin can be measured by differential-pulse polarography[115] or adsorptive stripping voltammetry,[116] with detection limits at the micromolar and subnanomolar concentration levels, respectively. The redox

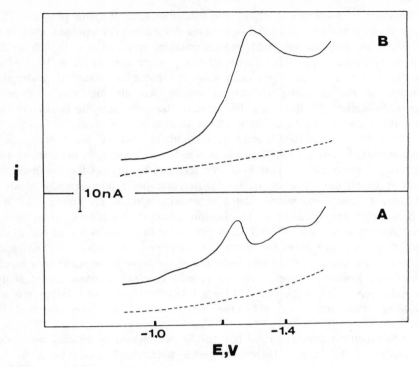

FIGURE 1-18
Adsorptive stripping voltammograms for $1 \times 10^{-8}\,M$
oxytetracycline (A) and doxycycline (B) following 5-min
preconcentration. The response without accumulation is shown as
dotted lines. (Reproduced with permission.[105])

reaction involves the reduction of the carbon-carbon double bond conjugated to the
carbonyl moiety. AC polarography in an aprotic organic solvent system and
adsorptive stripping voltammetry have been used to measure micromolar and
nanomolar levels, respectively, of the antihypertensive agent reserpine,[117, 118] while
adsorptive stripping voltammetry at carbon and mercury electrodes has been useful
for measuring trace levels of the calcium channel blockers nicardipine and
diltiazem, respectively.[119, 120]

The determination of other pharmaceutical compounds in body fluids has been
reported. Munson and Abdine[121] have described a rapid and simple differential-
pulse voltammetric procedure for the determination of acetaminophen in plasma.
The method is based on the oxidation of the phenolic moiety of actaminophen at a
carbon paste electrode. A well-defined differential-pulse peak is obtained, as shown
in Figure 1-20. The same authors reported a similar procedure for determining the
bronchodilator drug theophylline in plasma, following a simple solvent extraction
procedure.[122] Thioamide drugs, used as antiulcer agents, can be determined directly
in plasma and urine by cathodic stripping voltammetry.[123] The formation of

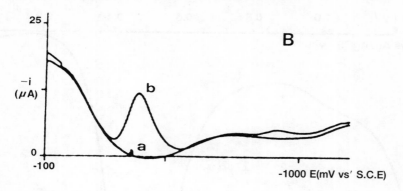

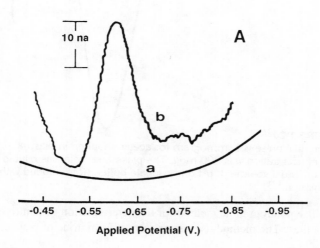

FIGURE 1-19
Differential-pulse polarograms for the anticancer drugs adriamycin
(A) and mitomycin C (B) in human plasma. (a) Plasma; (b) plasma
containing micromolar levels of the drugs. (Reproduced with
permission.[109, 112])

mercury(II)-sulfide at the hanging mercury drop electrode permits quantitation
down to the $2 \times 10^{-8}M$ level, without interference from naturally occurring
sulfur-containing compounds. Another antiulcer drug, cimetidine, can be determined in urine by adsorptive accumulation at the hanging mercury drop electrode,
followed by square-wave polarography of the surface species.[124] The polarographic
determination of phenobarbitol and diphenylhydantoin in blood has been reported
by Brooks et al.[125] The assay involves selective extraction into chloroform,
followed by controlled nitration of the drugs. The detection limit, $1-2$ μg/ml, is
similar to that of gas chromatographic assays. Cox et al.[126] have reported a

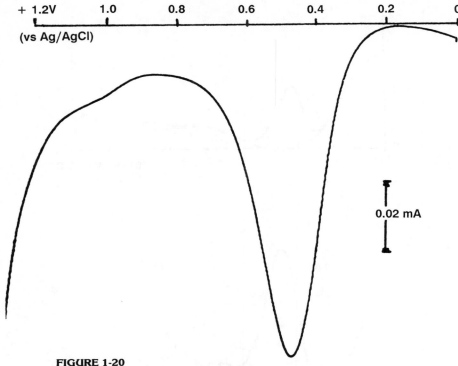

FIGURE 1-20
Differential-pulse voltammogram for acetaminophen in plasma.
The concentration was 200 mg/l. The plasma sample was diluted
with an equal volume of pH 8 phosphate buffer. (Reproduced with
permission.[121])

differential-pulse polarographic method for the determination of dantrolene sodium
and its metabolites. The method uses an extraction from urine or plasma with ethyl
acetate.

Various vitamins undergo reduction or oxidation processes and thus can be
determined at mercury or solid electrodes, respectively. For example, differential-
pulse polarography following solvent extraction can be used to monitor the
clearance of vitamin K_1 from human plasma after a 20-mg intravenous injection.[127]
Typical polarograms obtained for several plasma samples are shown in Figure 1-21.
A similar procedure for determining vitamin K (menadione) in plasma has been
described by Akman et al.[128] Polarography has been used to study the degradation
of vitamin K_1,[129] and the polarographic behavior of vitamin K_3 has been
reported.[130] Ascorbic acid undergoes an irreversible oxidation, to dehydroascorbic
acid, at low potentials. Such a process makes this vitamin suited to voltammetric
measurements at various solid electrodes.[131] Trace levels of vitamin B_2 (riboflavin)
can be measured by a highly sensitive adsorptive stripping procedure.[25] The
differential-pulse scan, following a 30-min preconcentration, yields a detection
limit of 2.5×10^{-11} M. Measurements of vitamin B_6 (pyridoxine), based on its

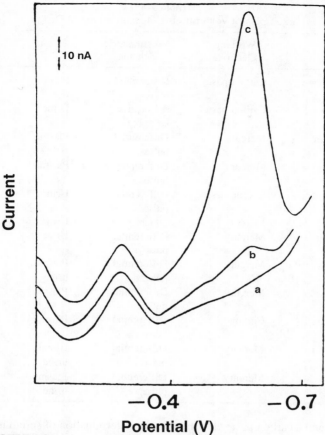

FIGURE 1-21
Differential-pulse polarograms of (a) control plasma; (b) plasma spiked with 412 ng of vitamin K_1, and (c) plasma spiked with 6.18 μg of vitamin K_1. (Reproduced with permission.[136])

well-defined anodic peak at carbon paste electrodes, have been described by Soderhjelm and Lindquist.[132] The same group reported similar procedures for vitamins A[133] and E.[134] These and other applications of voltammetric techniques to measurement of electroactive vitamins have been reviewed.[88, 135]

Selected applications of voltammetry for measuring drugs in clinical samples are given in Table 1-4.

1-13.3 Voltammetric Measurements of Naturally Occurring Biological Compounds

Uric acid is a major constituent of body fluids; abnormalities of uric acid levels in these fluids are symptoms of several diseases. At carbon electrodes, uric acid yields a well-defined oxidation peak at relatively low potentials. The use of fixed potential

Table 1-4
Examples of Voltammetric Measurements of Drugs

Drug	Working Electrode	Voltammetric Scheme	Matrix	Reference
Acetaminophen	Carbon paste	Differential pulse	Plasma	121
Adriamycin	Carbon paste	Adsorptive stripping	Urine	108
Bromazepam	Mercury	Differential pulse	Plasma	5
Chlordeaze-poxide	Mercury	Differential pulse	Plasma	93
Chlorpromazine	Carbon paste	Differential pulse	Urine	95
Cimetidine	Mercury	Square wave	Urine	124
cis-Platinum	Mercury	Differential pulse	Urine	106
Dontrolene	Mercury	Differential pulse	Plasma, urine	128
Mitomycin C	Mercury	Differential pulse	Plasma	112
Theophylline	Carbon	Differential pulse	Plasma	122
Trimethoprin	Mercury	Differential pulse	Blood, urine	102
Vinblastine	Mercury	Differential pulse	Plasma, urine	111

conditions and a carbon paste electrode for the determination of serum uric acid has been reported.[136] Linear scan[137] and adsorptive stripping[138] voltammetry can also be used for uric acid measurements.

Analytical measurements of nucleic acid bases are important in various situations. Cummings et al.[139] have described the utility of differential-pulse polarography for trace measurements of some adenine and cytosine species. The method yields linear calibration plots, with detection limits at the submicromolar concentration level. Trace levels of nucleic acid bases can be measured by means of cathodic stripping voltammetry.[140] Such a procedure is based on the formation of sparingly soluble compounds with mercury. The detection limit for adenine is about $2 \times 10^{-9}\ M$. The cathodic stripping and differential-pulse voltammetric behaviors of 30 purine and pyrimidine derivatives at the mercury electrode were evaluated by Palecek et al.[141] Applications of pulse voltammetry at carbon electrodes in nucleic acid analysis have been described by Brabec.[142] Differential-pulse voltammetry at a glassy carbon electrode can be used for the determination of 7-methylguanine.[143] This procedure provides a convenient appoach for in-vitro screening of the alkylation efficiency of alkylating agents exerting mutagenic and carcinogenic effects. Cyclic voltammetry following adsorptive accumulation allows the determination of submicrogram quantities of circular duplex DNA in plasmid samples.[144] Besides their clinical utility, voltammetric studies offer better understanding of the biological behavior of the genetic material. Palecek[145] has reviewed recent progress in voltammetric studies of the genetic material.

The important enzyme cofactor dihydronicotamide adenine dinucleotide (NADH) is oxidized at various solid electrodes. The oxidation process has been the subject of numerous investigations.[146] The technique of steady-state voltammetry at the rotating-disk electrode has been employed to obtain well-defined current-potential curves at the micromolar concentration level.[147] Various surface modification procedures have been employed to lower the overvoltage of the oxidation process.[148] Binding of the heme protein cytochrome C to the surface of various ordinary or modified electrodes facilitates its electron-transfer properties and allows its determination down to the micromolar concentration level.[149]

Measurements of bilirubin in body fluids is important for diagnostic purposes and therapeutic monitoring. Voltammetric methods for the determination of bilirubin can be based on either its oxidation at solid electrodes or its reduction at mercury electrodes. The latter is preferred due to sensitivity and reproducibility considerations. Differential-pulse[150] or fast square-wave[151] polarography and adsorptive stripping voltammetry[152] have been used for measurements of bilirubin at the micromolar and nanomolar concentration levels. In particular, the determination of total bilirubin in serum using the square-wave waveform yielded good correlation with an established spectrophotometric method.

Polarographic techniques have been applied for the measurement of unsaturated ketosteroids. In particular, the reduction of three unsaturated ketosteroids (where the carbonyl group is conjugated to the carbon-carbon double bond) has been exploited for measuring the compounds by ac and differential-pulse polarography[153, 154] or adsorptive stripping voltammetry[155] at the micromolar and nanomolar concentration levels, respectively. The sterols cholesterol and lynestrenol do not contain electroactive moieties but can be measured indirectly after chemical conversion to reducible compounds.[156, 157]

The biological significance of cysteine and cystine, coupled with their electrochemical activity, has prompted numerous voltammetric investigations of these compounds. Of clinical interest are differential-pulse[158] and cathodic stripping[159] voltammetric procedures. The latter, performed in the presence of cupric ions, results in a detection limit of $2 \times 10^{-9} M$. A similar approach can be applied for trace measurements of peptides and proteins containing disulfide linkages.[160] Disulfide-containing amino acids, peptides, and proteins also produce catalytic hydrogen polarographic prewaves (known as *Brdicka catalytic waves*) in the presence of cobalt or nickel.[161] Because of their catalytic nature, such waves yield more sensitive results than diffusion-controlled currents, often coupled with a nonlinear concentration dependence. A differential-pulse assay for total serum protein was reported recently.[162] The cellular constituents of whole blood did not interfere.

One of the most important applications of voltammetric techniques is the determination of neurotransmitters in brain tissues. These compounds are essential participants in the neurotransmission process and thus are implicated in various neurological diseases. The facile oxidation of neurotransmitters (to the corresponding o-quinones) provides the basis for their quantitation by differential-pulse[163] or normal-pulse[164] voltammetry at various solid electrodes, down to the submicromolar concentration level. Sternson et al.[165] reported a detailed study of the redox behavior of these compounds. Besides in-vitro studies, this behavior has been successfully utilized for the in-vivo monitoring of neurotransmitters released in the brains of small animals (see Chapter 5) and for their amperometric detection following separation by liquid chromatography (see Chapter 4).

1-13.4   Voltammetric Immunoassay

Immunoassay techniques are of great importance for specific and sensitive measurements of drugs, hormones, and other compounds of clinical significance. Hence, a wide variety of medical conditions can be diagnosed immunochemically. Such techniques are based on the competition for antibody (Ab) binding sites between a labeled (Ag) and unlabeled (Ag*) antigen:

$$
\begin{array}{ccc}
\text{Ag} & & \text{Ag:Ab} \\
+ & \rightleftharpoons & \\
\text{Ab} & & \\
+ & \rightleftharpoons & \\
\text{Ag*} & & \text{Ag*:Ab}
\end{array}
\qquad (1\text{-}17)
$$

Immunochemical reactions are among the most selective reactions known. They are based on shape recognition of the antigen by the antibody binding site. The high selectivity of immunochemical reactions thus allows assays in complex clinical media, with minimal interference. The possibility of immunological methods based on labeling an antigen with an electroactive functionality has been explored recently in several laboratories. These procedures are part of a new generation of immunological tests that do not require the use of radiolabeled materials. A voltammetric immunoassay relies on monitoring the binding via the decrease in the current response for the redox reaction of the labeled antigen in the presence of antibody. A similar principle applies to the labeled antibody current that decreases in the presence of antigen. This approach combines the inherent specificity of an antibody-antigen reaction with the sensitivity and wide linear range of advanced voltammetric schemes, e.g., differential-pulse or stripping voltammetry. Various disadvantages characterizing the use of isotopic labels (e.g., cost, radioactive waste) can thus be minimized or eliminated. Voltammetric immunoassay can be implemented in various ways, depending on the type of label, assay format, and voltammetric method. The assay format can be a heterogeneous one, requiring a separation step during the analytical procedure, or the more desirable homogeneous assay, involving no separation step. Electrode fouling from matrix constituents may be a problem in homogeneous assays.

In the first report on voltammetric immunoassay, Heineman et al.[166] labeled estriol with mercuric acetate (as the electroactive moiety) and monitored the reaction of this labeled antigen with estriol antibody. Separation of free-labeled antigen from antibody-bound-labeled antigen was unnecessary. Figure 1-22 shows a typical labeled antigen-antibody binding curve for 4-mercuric acetate estriol. Estriol can be made electroactive also by labeling with nitro groups in the 2 and 4 positions for an immunoassay with detection by differential-pulse polarography.[167] Weber and Purdy[168] described a homogeneous voltammetric immunoassay for morphine with ferrocene as the electroactive tag; detection was at a glassy carbon electrode in a flow-through cell. Christian's group investigated homogeneous immunoassays for human serum albumin labeled with various metal ions[169-171]; bound metal ions were detected by differential-pulse polarography. Metal ion labels can be employed also for heterogeneous assays. For example, serum proteins were assayed in conjunction with anodic stripping detection.[172] The assay involved covalently linking a chelating agent, such as diethyenetriaminepentaacetic acid, to the protein to serve as a chelon for an indium label. The latter was released by acidification (after competitive equilibrium) and was then measured using a differential-pulse stripping waveform. Indium was

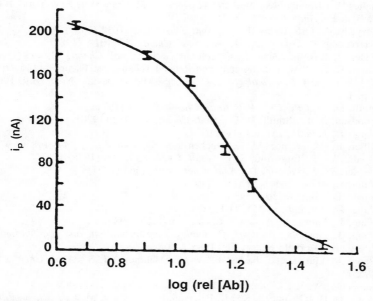

FIGURE 1-22
Labeled antigen-antibody binding curve for 4-mercuric acetate estriol with estriol antibody. (Reproduced with permission.[166])

chosen as the metal ion label because of its relative absence in clinical samples, its high-stability constant with the chelating agent, and its well-defined stripping behavior.

It is apparent, based on the above studies, that the use of electroactive labels has the potential to replace radioimmunoassay precedures for various clinical applications. Other strategies for electrochemical immunoassay, based on enzyme-labeled antigens and immunoelectrochemical sensors, are described in Chapters 3 and 4. The reader is referred to a recent review article[173] for a discussion of electrochemical immunoassay procedures.

REFERENCES

1. Adams, R. N., *Electrochemistry at Solid Electrodes*, Marcel Dekker, New York, 1969.
2. Bard, A. J., Faulkner, L. R., *Electrochemical Methods*, Wiley, New York, 1980.
3. Bond, A. M. *Modern Polarographic Methods in Analytical Chemistry*, Marcel Dekker, New York, 1980.
4. Kissinger, P. T., Heineman, W. R. (eds.), *Laboratory Techniques in Electroanalytical Chemistry*, Marcel Dekker, New York, 1984.
5. Brooks, M. A., Hackman, M. R., *Anal. Chem.*, *47*, 2059 (1975).
6. DeAngelis, T. P., Bond, R. E., Brooks, E. E., Heineman, W. R., *Anal. Chem.*, *49*, 1792 (1977).
7. Wang, J., Freiha, B. A., *Anal. Chem.*, *54*, 334 (1982).
8. Meites, L., *Handbook of Analytical Chemistry*, McGraw-Hill, New York, 1963.
9. Flato, J. B., *Anal. Chem.*, *44*, 75A (1972).
10. Parry, E. P., Osteryoung, R. A., *Anal. Chem.*, *37*, 1634 (1964).
11. Osteryoung, J., Osteryoung, R. A., *Anal. Chem.*, *57*, 101A (1985).
12. O'Dea, J. J., Osteryoung, J., Osteryoung, R. A., *Anal. Chem.*, *53*, 695 (1981).
13. Smith, D. E., "AC Polarography and Related Techniques: Theory and Practice," in A. J. Bard (ed.), *Electroanalytical Chemistry*, Marcel Dekker, New York, 1966, vol. 1, p. 1.
14. Smith, D. E., *CRC Crit. Rev. Anal. Chem.*, *2*, 247 (1971).
15. Woodson, A. L., Smith, D. E., *Anal. Chem.*, *42*, 242 (1970).
16. Wang, J., *Talanta*, *28*, 369 (1981).
17. Miller, B., Bellavance, M. I., Bruckenstein, S., *Anal. Chem.*, *44*, 1983 (1972).
18. Blaedel, W. J., Engstrom, R. C., *Anal. Chem.*, *50*, 476 (1978).
19. Blaedel, W. J., Iverson, D., *Anal. Chem.*, *49*, 1563 (1977).
20. Wang, J., *Anal. Chem.*, *53*, 1528 (1981).
21. Jagner, D., *Analyst*, *107*, 593 (1982).
22. Wang, J., *Am. Lab.*, *17*(5), 41 (1985).
23. Wang, J., Freiha, B. A., *Anal. Chem.*, *55*, 1285 (1983).
24. Wang, J., *Stripping Analysis: Principles, Instrumentation, and Applications*, VCH Publishers, Deerfield Beach, FL, 1985.
25. Wang, J., Luo, D. B., Farias, P. A. M., Mahmoud, J. S., *Anal. Chem.*, *57*, 158 (1985).
26. Heineman, W. R., Kissinger, P. T., *Am. Lab.*, *14*(11), 29 (1982).
27. Evans, D. H., O'Connell, K. M., Peterson, R. A., Kelly, M. J., *J. Chem. Educ.*, *60*, 290 (1983).
28. Sailasuta, N., Anson, F. C., Gray, H. B., *J. Am. Chem. Soc.*, *101*, 455 (1979).
29. Bancroft, E. E., Blount, H. N., Hawkridge, F. M., *Biochem. Biophys. Res. Commun.*, *101*, 1331 (1981).

30. Owens, J. L., Dryhurst, G., *J. Electroanal. Chem., 80,* 171 (1977).
31. Watanabe, T., Honda, K., *J. Am. Chem. Soc., 102,* 370 (1980).
32. Mayausky, J. S., McCreery, R. L., *Anal. Chem., 55,* 308 (1983).
33. Kuwana, T., Heineman, W. R., *Acc. Chem. Res., 9,* 241 (1976).
34. Heineman, W. R., *J. Chem. Educ., 60,* 305 (1983).
35. Moros, S. A., Hamilton, C. M., Heveran, J. E., Donahue, J. J., Oliveri-Vigh. *J. Pharm. Sci.,* 1229 (1975).
36. Girard, M. L., Rousselet, F., Fouye, H., Levillain, P., *Ann. Pharm. Fr., 27,* 173 (1967).
37. Patriarche, G., *Contribution a l'Analyse Coulometrique: Applications aux Sciences Pharmaceutiques,* Editions Arscia S.A., Brussels, 1964.
38. Harrar, J. E., "Techniques, Apparatus, and Analytical Applications of Controlled-Potential Coulometry," in A. J. Bard (ed.), *Electroanalytical Chemistry,* Marcel Dekker, New York, 1974, vol. 8, p. 1.
39. Jacobsen, E., Jacobsen, T. V., *Anal. Chim. Acta, 55,* 293 (1971).
40. Patriarche, G., Lingane, J. J., *Anal. Chim. Acta, 49,* 241 (1970).
41. Zuman, P., *The Elucidation of Organic Electrode Processes,* Academic Press, New York, 1969.
42. Lund, W., Eriksen, R., *Anal. Chim. Acta, 107,* 37 (1979).
43. Newberry, C. L., Christian, G. D., *J. Electroanal. Chem., 9,* 468 (1965).
44. Adeloju, S. B., Bond, A. M., Noble, M. L., *Anal. Chim. Acta, 161,* 303 (1984).
45. Adeloju, S. B., Bond, A. M., Briggs, M. H., *Anal. Chem., 56,* 2397 (1984).
46. Morrell, G., Giridhar, G., *Clin. Chem., 22,* 221 (1976).
47. Bond, A. M., Reust, J. B., *Anal. Chim. Acta, 162,* 389 (1984).
48. Simon, R. K., Christian, G. D., Purdy, W. C., *Am. J. Clin. Phatol., 49,* 733 (1986).
49. Veillon, C., *Anal. Chem., 58,* 851A (1986).
50. Nürnberg, H. W., *Pure Appl. Chem., 54,* 853 (1982).
51. Adeloju, S. B., Bond, A. M., Briggs, M. H., *Anal. Chem., 57,* 1386 (1985).
52. Franke, J. P., de Zeeuw, R. A., *Arch. Toxicol., 37,* 47 (1976).
53. Nürnberg, H. W., *Electrochim. Acta, 22,* 935 (1977).
54. Peter, F., Reynolds, R. G., *Anal. Chem., 48,* 2041 (1976).
55. Levit, D. I., *Anal. Chem., 45,* 1291 (1973).
56. Jagner, D., Josefson, M., Westerlund, S., Aren, K., *Anal. Chem., 53,* 1406 (1981).
57. Wang, J., *Anal. Chem., 54,* 221 (1982).
58. Chittleborough, G., Steel, B. J., *Anal. Chim. Acta, 119,* 235 (1980).
59. Oehme, M., Lund, W., Jonsen, J., *Anal. Chim. Acta, 100,* 389 (1978).
60. Williams, T. R., Foy, O. P., Benson C., *Anal. Chim. Acta, 75,* 250 (1975).
61. Hutin, M. F., Netter, P., Burnell, D., Gaucher, A., *Rhumatologie, 12,* 199 (1985).
62. Boone, J., Hearn, T., Lewis, S., *Clin. Chem., 25,* 389 (1979).
63. Wang, J., *J. Electroanal. Chem., 139,* 225 (1982).
64. Duic, L., Szechter, S., Srinivasan, S., *J. Electroanal. Chem., 41,* 83 (1973).
65. Golimowski, J., Valenta, P., Stoeppler, M., Nürnberg, H. W., *Talanta, 26,* 649 (1979).
66. Golimowski, J., Valenta, P., Stoeppler, M., Nürnberg, H. W., *Z. Anal. Chem., 290,* 107 (1978).
67. Valenta, P., Rutzel, H., Nürnberg, H. W., Stoeppler, M., *Z Anal. Chem., 284,* 1 (1977).
68. Oehme, M., Lund, W., *Z. Anal. Chem., 298,* 260 (1979).
69. Copeland, T. R., Christie, J. H., Osteryoung, R. A., Skogerboe, R. K., *Anal. Chem., 45,* 2171 (1973).
70. Davis, P., Berlandi, F., Dulude, G., Griffin, R., Matson, W., *Am. Ind. Hyg. Assoc. J., 6,* 480 (1978).
71. Kinard, J. T., *Anal. Lett., 10,* 1147 (1977).
72. Attramattal, A., Jonsen, J., *Acta Odontol. Scand., 34,* 127 (1976).

73. Franke, J. P., Coenegrach, P. M. J., de Zeeuw, R. A., *Arch. Toxicol.*, *34*, 137 (1975).
74. Constantine, S., Giordano, R., Rizzica, M., Benedetti, F., *Analyst*, *110*, 1355 (1985).
75. Adeloju, S. B., Bond, A. M., Briggs, M. H., *Anal. Chim. Acta*, *164*, 181 (1984).
76. Gammelgaard, B., Andersen, J. R., *Analyst*, *110*, 1197 (1985).
77. Schmid, G. M., Bolger, G. W., *Clin. Chem.*, *19*, 1002 (1973).
78. Coetzee, J. F., Hussan, A., Petrick, T. R., *Anal. Chem.*, *55*, 120 (1983).
79. Jagner, D., Aren, K., *Anal. Chim. Acta*, *141*, 157 (1982).
80. Kauffman, J. M., Patriarche, G. J., Christian, G., *Anal. Lett.*, *14*, 1409 (1981).
81. Jagner, D., Danielsson, L. G., Aren, K., *Anal. Chim. Acta*, *106*, 15 (1979).
82. Kemula, W., Kublik, Z., *Nature* (London), *189*, 57 (1961).
83. Ferren, W. P., *Am. Lab.*, *7*, 52 (1978).
84. Schreiber, M. A., Last, T. A., *Anal. Chem. 53*, 2095 (1981).
85. Adeloju, S. B., Bond, A. M., Hughes, H. C., *Anal. Chim. Acta*, *148*, 59 (1983).
86. Simon, J., Liese, T., *Z. Anal. Chem.*, *314*, 483 (1983).
87. Brooks, M. A., *Bioelectrochem. Bioenerg.*, *10*, 37 (1983).
88. Patriarche, G. J., Chateau-Gosselin, M., Vandenbalck, Zuman, P., "Polarography and Related Electroanalytical Techniques in Pharmacy and Pharmacology," in A. J. Bard (ed.), *Electroanalytical Chemistry*, Marcel Dekker, New York, 1979, vol. 11, p. 141.
89. Smyth, W. F., *Polarography of Molecules of Biological Significance*, Academic Press, London, 1979.
90. Volke, J., *Bioelectrochem. Bioenerg.*, *10*, 17 (1983).
91. Smyth, M. R., Smyth, W. R., *Analyst*, *103*, 529 (1978).
92. de Silva, J. A. F., Bekersky, I., Brooks, M. A., Weinfeld, R. E., Glover, W., Puglisi, C. V., *J. Pharm. Sci.*, *63*, 1440 (1974).
93. Hackman, M. R., Brooks, M. A., de Silva, J. A. F., Ma, T. S., *Anal. Chem.*, *45*, 263 (1973).
94. Kalvoda, R., *Anal. Chim. Acta*, *162*, 197 (1984).
95. Jarbawi, T. B., Heineman, W. R., Patriarche, G. J., *Anal. Chim. Acta*, *57*, 126 (1981).
96. Wang, J., Freiha, B. A., Deshmukh, B. K., *Bioelectrochem. Bioenerg.*, *14*, 457 (1985).
97. Jarbawi, T. B., Heineman, W. R., *Anal. Chim. Acta*, *186*, 11 (1986).
98. Wang, J., Bonakdar, M., Morgan, C., *Anal. Chem.*, *58*, 1024 (1986).
99. Oelschlager, H., *Bioelectrochem. Bioenerg.*, *10*, 25 (1983).
100. Siegerman, H., "Polarography of Antibiotics and Antibacterial Agents," in A. J. Bard (ed.), *Electroanalytical Chemistry*, Marcel Dekker, New York, 1979, vol. 11, p. 291.
101. Wang, J., Mahmoud, J. S., *Anal. Chim. Acta*, *186*, 31 (1986).
102. Brooks, M. A., de Silva, J. A. F., D'Arconte, L. M., *Anal. Chem.*, *45*, 263 (1973).
103. Forsman, U., *Anal. Chim. Acta*, *146*, 71 (1983).
104. Ivaska, A., Nordstrom, F., *Anal. Chim. Acta*, *146*, 87 (1983).
105. Wang, J, Tuzhi, P., Lin, M. S., *Bioelectrochem. Bioenerg.*, *15*, 147 (1986).
106. Vrana, O., Kleinwachter, V., Brabee, V., *Talanta*, *30*, 288 (1983).
107. Schmid, G. M., Atherton, D. R., *Anal. Chem.*, *58*, 1956 (1986).
108. Chaney, E. N., Baldwin, R. P., *Anal. Chem.*, *54*, 2556 (1982).
109. Sternson, L. A., Thomas, G., *Anal. Lett.*, *10*, 99 (1977).
110. Rusling, J. F., Scheer, B. J., Haque, I. U., *Anal. Chim. Acta*, *158*, 23 (1984).
111. Temizer, A., *Talanta*, *33*, 791 (1986).
112. Van Bennekom, W. P., Tjaden, U. R., De Bruijn, E. A., Van Oosterom, A. T., *Anal. Chim. Acta*, *156*, 289 (1984).
113. Wang, J., Tuzhi, P., Lin, M. S., Tapia, T., *Talanta*, *33*, 707 (1986).
114. Wang, J., Lin, M. S., Villa, V., *Anal. Lett.*, *19*, 2293 (1986).
115. Kadish, K. M., Spiehler, V. R., *Anal. Chem.*, *47*, 1714 (1975).
116. Wang, J., Mahmoud, J. S., Farias, P. A. M., *Analyst*, *110*, 855 (1985).
117. Taira, A., Smith, D. E., *J. Assoc. Off. Anal. Chem.*, *61*, 941 (1978).

118. Wang, J., Tapia, T., Bonakdar, M., *Analyst, 111,* 1245 (1986).
119. Wang, J., Deshmukh, B. K., Bonakdar, M., *Anal. Lett., 18,* 1087 (1985).
120. Wang, J., Farias, P. A. M., Mahmoud, J. S., *Analyst, 111,* 837 (1986).
121. Munson, J. W., Abdine, H., *J. Pharm. Sci., 67,* 1775 (1978).
122. Munson, J. W., Abdine, H., *Talanta, 25,* 221 (1978).
123. Davidson, I. E., Smyth, W. F., *Anal. Chem., 49,* 1195 (1977).
124. Webber, A., Shah, M., Osteryoung, J., *Anal. Chim. Acta, 154,* 105 (1983).
125. Brooks, M. A., de Silva, J. A. F., Hackman, M. R., *Anal. Chim. Acta, 64,* 165 (1973).
126. Cox, P. L., Heotis, J. P., Polin, D., Rose, G. M., *J. Pharm. Sci., 58,* 987 (1969).
127. Hart, J. P., Nahir, A. M., Chayen, J., Catterall, A., *Anal. Chim. Acta, 144,* 267 (1982).
128. Akman, S. A., Kusu, F., Takamura, K. T., Chlebowski, R., Block, J., *Anal. Biochem., 141,* 488 (1984).
129. Vire, J. C., Patriarche, G. J., Christian, G. D., *Pharmazie 35,* 209 (1980).
130. Vire, J. C., Patriarche, G. J., Christian, G. D., *Anal. Chem., 51,* 752 (1979).
131. Lindquist, J., *Analyst, 100,* 339 (1975).
132. Soderhjelm, P., Lindquist, J., *Analyst, 100,* 349 (1975).
133. Atuma, S., Lindquist, J., Lundstrom, K., *Analyst, 99,* 683 (1974).
134. Atuma, S., Lindquist, J., *Analyst, 98,* 884 (1973).
135. Zuman, P., Brezina, M., in P. Zuman and I. M. Kolthoff (eds.), *Progress in Polarography,* Wiley-Interscience, New York, 1966, vol. 2, p. 687.
136. Park, G., Adams, R. N., White, W. R., *Anal. Lett., 5,* 887 (1972).
137. Dryhurst, G., De, P. K., *Anal. Chim. Acta, 58,* 183 (1972).
138. Wang, J., Freiha, B. A., *Bioelectrochem. Bioenerg., 12,* 225 (1984).
139. Cummings, T. E., Fraser, J. R., Elving, P. J., *Anal. Chem., 52,* 558 (1980).
140. Palecek, E., *Anal. Lett., 13,* 331 (1980).
141. Palecek, E., Jelen, F., Hung, M. A., *Bioelectrochem. Bioenerg., 8,* 621 (1981).
142. Brabec, V., *Bioelectrochem. Bioenerg., 8,* 437 (1981).
143. Sequaris, I. M., Valenta, P., Nürnberg, H. W., *J. Electroanal. Chem., 122,* 263 (1981).
144. Boublikova, P., Vojtiskova, M., Palecek, E., *Anal. Lett., 20,* 275 (1987).
145. Palecek, E., *Bioelectrochem. Bioenerg., 15,* 275 (1986).
146. Moiroux, J., Elving, P. J., *Anal. Chem., 51,* 346 (1979).
147. Blaedel, W. J., Jenkins, R. A., *Anal. Chem., 47,* 1337 (1975).
148. Jaegfeldt, H., Torstensson, A. B. C., Gorton, L. G. O., Johansson, G., *Anal. Chem., 53,* 1979 (1981).
149. Wang, J., Lin. M. S., *J. Electroanal. Chem., 221,* 257 (1987).
150. Slifstein, C., Ariel, M., *J. Electroanal. Chem., 75,* 551 (1977).
151. Saar, J., Yarnitzky, C., Israel, *J. Chem., 23,* 249 (1983).
152. Wang, J., Luo, D. B., Farias, P. A. M., *J. Electroanal. Chem., 185,* 61 (1985).
153. Woodson, A. L., Smith, D. E., *Anal. Chem., 42,* 242 (1970).
154. Chatten, L. G., Yadav, R. N., Madan, D. K., *Pharm. Acta Helv., 51,* 381 (1976).
155. Wang, J., Farias, P. A. M., Mahmoud, J. S., *Anal. Chim. Acta, 171,* 195 (1985).
156. Furst, *Pharm. Zentralh, 107,* 184 (1967).
157. Van Bennekom, W. P., Reevwijk, J. E. M., Schute, J. B., *Anal. Chim. Acta, 74,* 387 (1975).
158. Mairesse-Ducarmois, C. A., Patriarche, G. J., Vandenbalk, J. L., *Anal. Chim. Acta, 71,* 165 (1974).
159. Forsman, V., *J. Electroanal. Chem, 122,* 215 (1981).
160. Forsman, V., *Anal. Chim. Acta, 166,* 141 (1984).
161. Banica, F. G., *Talanta, 32,* 1145 (1985).
162. Hertl, W., *Anal. Biochem., 164,* 1 (1987).
163. Lane, R. F., Hubbard, A. T., *Anal. Chem., 48,* 1287 (1976).

164. Ponchon, J. L., Cespuglio, R., Jouvet, M., Pujol, J. F., *Anal. Chem., 51,* 1483 (1978).
165. Sternson, A. W., McCreery, R., Feinberg, B., Adams, R. N., *J. Electroanal. Chem., 46,* 313 (1973).
166. Heineman, W. R., Anderson, C. W., Halsall, H. B., *Science, 204,* 865 (1979).
167. Wehmeyer, K. R., Halsall, H. B., Heineman, W. R., *Clin. Chem, 28,* 1968 (1982).
168. Weber, S. G., Purdy, W. C., *Anal. Lett., 12,* 1 (1979).
169. Alam, I. A., Christian, G. D., *Anal. Lett., 15,* 1449 (1979).
170. Alam, I. A., Christian, G. D., *Z. Anal. Chem., 320,* 281 (1985).
171. Alam, I. A., Christian, G. D., *Z. Anal. Chem, 318,* 33 (1984).
172. Doyle, M. J., Halsall, H. B., Heineman, W. R., *Anal. Chem., 54,* 2318 (1982).
173. Heineman, W. R., Halsall, H. B., *Anal. Chem., 57,* 1321A (1985).

CHAPTER 2 ——————————————————

Ion-Selective Electrodes in Clinical Medicine

2-1 PRINCIPLES OF POTENTIOMETRIC MEASUREMENTS

In potentiometry, information on the composition of a sample is obtained through the potential appearing between two electrodes. Potentiometry is a classical analytical technique with roots before the turn of this century. However, the rapid development of new electrodes and more sensitive and stable electronic components in the past 20 years has expanded tremendously the range of analytical applications of potentiometric measurements. The speed at which this field has developed is a measure of the degree to which potentiometric measurements meet the need of the clinical chemist for rapid, low-cost, and accurate analysis. In this chapter, the principles and clinical utility of direct potentiometric measurements—based on ion-selective electrodes—will be described. (The second major part of potentiometry, the so-called potentiometric titrations, will not be covered.)

The equipment required for direct potentiometric measurements includes an ion-selective electrode, a reference electrode, and a potential measuring device (a pH/millivolt meter that can read to 0.2 mV or better). The ion-selective electrode is an indicator electrode that preferentially responds to one ionic species. Such electrodes are mainly membrane-based devices, consisting of permselective ion-conducting materials, which separate the sample from the inside of the electrode. On the inside is a solution containing the ion of interest at a constant activity. The composition of the membrane is designed to yield a potential that is primarily due to the ion of interest (via selective binding processes, e.g., ion exchange, which occur at the membrane-solution interface). The trick is to find a membrane that will selectively bind the analyte ions, leaving co-ions behind. Detailed theory of the processes at the membrane interface, which generate the potential, is available elsewhere.[1-3] Thermodynamic arguments, which will not be elaborated here, tell us that the junction potential produced at the membrane corresponds to the free energy difference (ΔG) associated with the gradient of activity (of the analyte ions in the outer and inner solutions). The resulting potential of the ion-selective electrode, which reflects the unequal distribution of the analyte ions across the boundary, is generally monitored relative to the potential of a reference electrode. Since the potential of the reference electrode is fixed, the measured cell potential can be related to the activity of the dissolved ion.

49

Ion-selective electrodes should ideally obey Equation 2-1:

$$E = K + (2.303 \, RT/ZF)\log a_i \tag{2-1}$$

where E is the potential, R the gas constant (8.314 J/Kmol), T the absolute temperature, Z the ionic charge, and a_i the activity of ion i. K is a constant, containing contributions from various sources, e.g., several liquid junction potentials. Equation 2-1 predicts that the electrode potential is proportional to the logarithm of the activity of the ion monitored. For example, at room temperature a 59.1-mV change in the electrode potential should result from a tenfold change in the activity of a monovalent ion ($Z = 1$). Similar changes in the activity of a divalent ion should result in a 29.6-mV change of the potential. A 1-mV change in the potential corresponds to 4% and 8% changes in the activity of monovalent and divalent ions, respectively. The activity of an ion i in solution is related to its concentration, c_i, by the equation:

$$a_i = \gamma_i c_i \tag{2-2}$$

where γ_i is the activity coefficient. The activity coefficient depends on the types of ions present and on the total ionic strength of the solution.

Equation 2-1 has been written on the assumption that the electrode responds only to the ion i. In practice, ion-selective electrodes often exhibit some response to other ions. The response of the electrode in the presence of an interfering ion j is given by the Nikolskii-Eisenman equation:

$$E = K + (2.303 \, RT/Z_iF) \log(a_i + k_{ij}a_j^{\,Z_i/Z_j}) \tag{2-3}$$

where Z_i and Z_j are charges of ions i and j, respectively, and k_{ij} is the selectivity coefficient, a quantitative measure of the electrode ability to discriminate against the interfering ion (i.e., a measure of the relative affinity of ions i and j toward the ion-selective membrane). For example, if an electrode is 100 times more responsive to i than to j, k_{ij} has a value of 0.01. Ideally, the selectivity coefficient should be very small; selectivity coefficients lower than 10^{-5} have been achieved for several electrodes. It is important for the clinical chemist to realize the selectivity coefficient of a particular electrode in use based on Equation 2-3. Obviously, the error in the activity a_i due to the interference of j would depend upon their relative levels. Tables for the required selectivity coefficients of various ion-selective electrodes (for a maximally tolerable error), assuming representative physiological levels, are available.[4]

Usually, the clinical chemist has need to determine the concentration of the ion of interest rather than its activity. The obvious approach to convert potentiometric measurements from activity to concentration is to make use of an empirical calibration curve, such as the one shown in Figure 2-1. Electrode potentials of standard solutions are measured and plotted (on a semilog paper) versus the

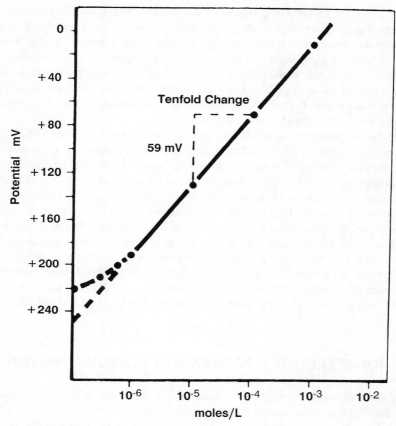

FIGURE 2-1
Typical calibration plot for a monovalent ion.

concentration. Theoretically, such a plot should yield a straight line, with a slope of approximately 59/Z mV (Nernstian slope). Detection by means of ion-selective electrodes may be performed over an exceedingly broad concentration range, which, for certain electrodes, may embrace five orders of magnitude. In practice, the usable range depends on other ions in the solution. Departure from the linearity is commonly observed at low concentrations (about 10^{-5}–10^{-6} M) due to the presence of co-ions (Equation 2-3). Similarly, at high concentrations of the ions of interest, interference by species of opposite charge (not described by Equation 2-3) may lead to deviation from the linear electrode response. Since the ionic strength of the sample is usually unknown, it is often useful to add a high concentration of an electrolyte to the standards and the sample to maintain about the same ionic strength. Hence, the empirical calibration plot yields results in terms of concentration.

Differences in the composition of the sample and standard solutions may account also for errors due to the formation of different liquid junction potentials between the reference electrode and the solution. One approach to alleviate this problem is to use an intermediate salt bridge, with a solution (in the bridge) of ions of nearly equal mobility. Standard solutions with an electrolyte composition similar to the sample are also desirable. These precautions, however, will not eliminate the problem completely. For example, erythrocyte suspensions are known to influence the liquid junction in whole blood analysis.

Different strategies for producing an electrode that is selective to one species have been suggested. These are based primarily on the nature and composition of the membrane material. The various membrane materials can thus be classified into the following categories: glass membranes, solid membranes (based on various crystalline materials), and liquid membranes (with a neutral carrier or ion-exchanger). Electrodes based on these membranes are produced by firms such as Orion Research, Beckman, Corning Glass Works, Phillips, Radiometer, Fisher Scientific, Graphic Control, Radelkis, Lazar Electrodes, and Sensorex. Size reduction, needed for assaying small sample volumes or for in-vivo monitoring, can be achieved by removal of the internal reference solution, as with coated-wire electrodes or ion-sensitive field effect transistors. In the following sections we will explore some ion-selective electrodes commonly used in clinical medicine.

2-2   ION-SELECTIVE ELECTRODES IN CLINICAL ANALYSIS

Body fluids are complex mixtures containing a wide variety of diverse substances. The determination of the ionic composition of these fluids is essential for establishing a patient's metabolic condition and for facilitating the diagnosis of altered states. The use of ion-selective electrodes in clinical medicine has grown rapidly in recent years. The determination of cationic electrolytes (e.g., Na^+, K^+, H^+, or Ca^{+2}), anions (e.g., Cl^-), and gases (e.g., CO_2, NH_3) in body fluids is now routine. Numerous commercial ion-selective electrode-based instruments are on the market, and new ones appear each month. Indeed, most new analyzers used in large hospitals now employ ion-selective electrode technology (Table 2-1). For example, potentiometric measurements of potassium and sodium have received widespread clinical acceptance, frequently replacing flame photometry for routine analysis. Ion-selective electrodes offer various advantages to the clinical chemist. They are low in cost, simple, and easily automated; they require minimal additional reagents or sample pretreatment (as neither turbidity nor color of the sample is a problem), and they can provide rapid assays of clinical samples (with volumes as small as a submicroliter). Hence, no flames or compressed gases are required as compared with atomic spectroscopic methods. An advantage or disadvantage, depending on the situation, is that ion-selective electrodes are responsive only to the free (uncomplexed) form of the element.

Table 2-1
Commercial Analyzers Based on Ion-Selective Electrodes

Analyzer	Company	Analytes
Abalyte	Abbot	Na, K
Astra	Beckman	Na, K
Lablyte-830	Beckman	Na, K, Li
AVL-980	AVL	Na, K, Ca
AVL-985	AVL	Na, K, Li
Corning-902	Corning	Na, K
Dupont ACA	Dupont	Na, K
Ektachem 400	Eastman Kodak	Na, K, Cl, CO_2
Hitachi 702	Hitachi	Na, K, Cl
ICA- I	Radiometer	Na, K, Ca
IL-502	Instrumentation Laboratory	Na, K
Lytening	Amdev	Na, K
Nova-4	Nova Biomedical	Na, K, Cl, CO_2
SS-20	Orion Research	Ca
SS-30	Orion Research	Na, K
Technicon RA-1000	Technicon	Na, K, CO_2

Some of the commercial instruments dilute the sample before the potentiometric measurement, while others do not. A schematic representation of diluted and undiluted sampling methods is given in Figure 2-2. There are tradeoffs in the use of direct versus indirect measurements. Direct potentiometry (measurements in undiluted samples) is attractive because no sample preparation is required, and the assay values are not sensitive to the levels of solids in clinical samples. Because direct potentiometry is unaffected by changes in plasma water, it permits accurate

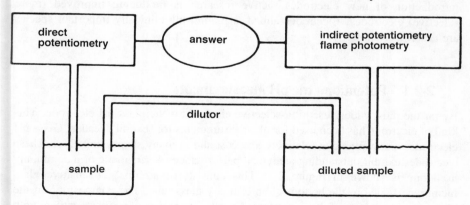

FIGURE 2-2
Schematic representation of undiluted and diluted sampling techniques used by direct and indirect potentiometry and flame photometry methods. (Reproduced with permission.[5])

measurement of electrolytes in patients with hyperlipemia or multiple myeloma. You must, however, be aware of junction potentials at the reference electrode and activity coefficients. Indirect potentiometry (dilution procedures) is attractive because it yields data equivalent to data obtained by flame photometry, and is well suited for use on high-throughput analyzers. The dilution step eliminates possible differences in the activity coefficient or ion binding from sample to sample. On the other hand, attention should be given to the total solids content of the sample. During the early years of commercialization, little distinction was made between direct and indirect methods, and this was a source of confusion in the clinical arena. Extensive research has resulted in better understanding of the differences between the two approaches. Nevertheless, standard methods are urgently needed for resolving the distinction between these approaches. It is now becoming clear that both methods yield data that are clinically useful but not necessarily the same.

Short- and long-term stability are important properties of ion-selective electrodes of great relevance to clinical measurements.[4] For example, a drift in potential often occurs after first contact with a protein-containing solution due to changes in the interfacial conditions (deposition of proteins). Surface contamination, membrane poisoning, loss of membrane components, and electrical leakage pathways can all affect the long-term stability. For example, the lipophilic nature of blood samples favors a substantial loss of liquid membrane components, such as an ion carrier. This represents a problem for applications involving continuous contact with blood samples (e.g., in-vivo monitoring). In commercial clinical analyzers, in contrast, this problem is alleviated by the periodic flow of the washing solution.

The number of ion-selective electrodes developed in the past 20 years has increased enormously, as illustrated in comprehensive reviews.[4, 6-10] Besides the introduction of new electrodes, active research is producing improved (more selective) electrodes. Electrodes aimed at monitoring clinically important species are described below.

2-2.1   Potentiometric pH Measurements

By far the most widely used ion-selective electrode is the glass pH electrode. This kind of electrode has been used for pH measurements for several decades. Glass pH electrodes are available at relatively low cost and in many shapes and sizes. These electrodes exhibit outstanding analytical performance. A schematic of a commonly used one is shown in Figure 2-3. This consists of a thin, pH-sensitive glass membrane sealed to the bottom of an ordinary glass tube. The composition of the glass membrane is carefully controlled. Usually, it consists of a silicate lattice, with negatively charged oxygen atoms, coordinating to cations such as sodium and calcium or (in modern formulations) to lithium and barium. Some of the more popular glasses have compositions of 72% SiO_2–22% Na_2O–6% CaO or 80% SiO_2–10% Li_2O–10% CaO. Inside the glass bulb is a dilute hydrochloric acid

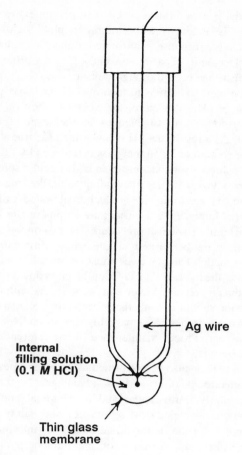

FIGURE 2-3
A glass pH electrode.

solution and a silver wire coated with a layer of silver chloride. The electrode is immersed in the solution whose pH is to be measured, and connected to an external reference electrode. (In the so-called combination electrode, the external reference electrode is combined with the ion-selective electrode into a single, integrated structure.) The rapid equilibrium established across the glass membrane, with respect to the hydrogen ions in the inner and outer solutions, produces a potential

$$E = K + (RT/F) \, ln \, (H^+)inner/(H^+)outer \qquad (2\text{-}4)$$

The potential of the electrode is registered with respect to the external reference electrode. Hence, the cell potential (at 25°C and after introducing the definition of pH) follows the relation

$$E_{cell} = K' + 0.059 \, pH \qquad (2\text{-}5)$$

The measured potential is thus a linear function of pH; an extremely wide (10–14 decades) linear range is obtained, with calibration plots yielding a slope of 59 mV/pH unit. The selective response is attributed to the ion-exchange properties of the glass surface (replacements of metal ions in the glass by protons). The theory of the response mechanism has been thoroughly discussed.[11]

The user must be alert to some shortcomings of the glass pH electrode. For example, in solutions of pH 10 or more, the electrode shows a so-called *alkaline error* in which it responds also to changes in the level of alkali metal ions (particularly sodium). As a result, the pH is lower than the true value. This error is greatly reduced if the sodium oxide in the glass is replaced by lithium oxide. Many glass electrodes also exhibit erroneous results in highly acidic solutions (pH < 0.5); the so-called *acid error* yields higher pH readings than the true value.

Before using the pH electrode, it should be calibrated using two (or more) buffers of known pH. Many standard buffers are commercially available, with an accuracy of ±0.01 pH unit. The exact procedure depends on the model of pH meter used. Calibration must be performed at the same temperature at which the measurement will be made. The electrode must be stored in an aqueous solution when not in use, so that the hydrated gel layer of the glass does not dry out. A highly stable response can thus be obtained over long periods. As with other ion-selective electrodes, the operator should consult the manufacturer's instructions for proper use. Commercial glass electrodes are remarkably robust, and with proper care, will last for more than a year. Proper maintenance of the reference electrode is also essential to minimize errors.

Measurements of pH can also be performed using other types of potentiometric sensors. Nonglass electrodes offer various advantages for certain clinical pH measurements (particularly intravascular and intraluminal applications), including ease of preparation, low electrical resistance, and safety in handling. The most common examples are the quinhydrone electrode (in which the response is due to a proton-transfer redox reaction of the quinone-hydroquinone couple) and the antimony electrode (based on the redox reaction between antimony and antimony oxide involving protons). Other metal–metal oxide couples, e.g., palladium–palladium oxide, have been applied for pH measurements. Membrane electrodes based on various neutral hydrogen ion carriers can also be employed. The neutral carrier tridodecylamine is especially useful for clinical use.[12] The resulting electrode exhibits excellent selectivity, reproducibility, and accuracy, but its dynamic range is inferior compared with glass electrodes. (Such a range appears to depend on the acidity constant of the incorporated ionophore.) New pH sensors based on new glass compositions or nonglass formulations are currently being developed in various laboratories. While such electrodes may be useful for specific applications, glass electrodes are likely to remain the choice for routine measurements with clinical analyzers.

Numerous biomedical applications of pH electrodes have been reported. Glass pH electrodes of different shapes and sizes have been particularly useful for these applications. The theoretical pH response (Equation 2-5) is commonly observed under physiological conditions. For pH measurements of blood, relevant to actual physiological situations, the body temperature, 37°C, should be used. For a discussion of the effect of temperature on the measurement, see Christian.[13] Venous blood is commonly used for pH measurements, with samples kept anaerobically to prevent changes in the level of carbon dioxide. Between measurements, the electrode should be rinsed with saline solution to prevent its coating. Maas et al.[14] described a reference method for accurate measurement of blood pH using the National Bureau of Standards buffers. It is not within the scope of this chapter to cover the vast number of clinical applications of pH electrodes. Miniature pH electrodes, designed mainly for in-vivo monitoring, are described in Chapter 5.

2-2.2 Potassium and Sodium Electrodes

In recent years, ion-selective electrodes have been increasingly used for the determination of potassium and sodium in body fluids. The growing popularity of K^+-Na^+ potentiometric measurements is illustrated in Figure 2-4. Numerous microprocessor-based K^+-Na^+ blood analyzers have appeared on the market, as indicated in Table 2-1. Some of the analyzers employ sample dilution, while others do not. Some controversy exists regarding the values obtained for sodium measurements on undiluted systems (see the discussion below). In addition, flame photometry has been employed for many years in clinical laboratories, yielding K^+-Na^+ values on a concentration basis. Although activity data may be more clinically appropriate, the usual concentration ranges are likely to be retained, because confusion in interpretation of the data by physicians is suspected. It has been proposed to convert the potentiometric activity data into total concentrations by multiplying the results by an appropriate factor.[16]

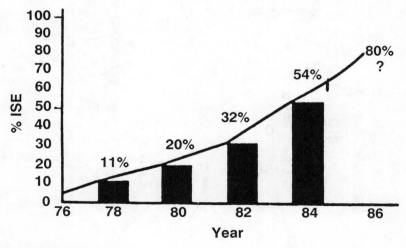

FIGURE 2-4
Percentage of total sodium and potassium hospital assays
performed with ion-selective electrodes during the last 10 years.
(Reproduced with permission.[15])

2-2.2.1 POTASSIUM MEASUREMENTS

The development of the valinomycin-based potassium-selective electrode represents one of the most significant events in the history of ion-selective electrodes. All modern clinical analyzers now rely on this electrode for potassium measurements.

Valinomycin is a cyclic dodecadepsipeptide, which contains 12 residues of alternating ester and peptide linkages (Figure 2- 5). This doughnut-shaped molecule has an electron-rich pocket in the center into which potassium ions are selectively extracted. For example, the electrode exhibits a selectivity for K^+ over Na^+ of approximately 5000. The basis for the selectivity seems to be the fit between the size of the potassium ion (radius 1.33 Å) and the volume of the cavity of the macrocyclic molecule. In addition to its excellent selectivity, such an electrode is well-behaved and has a wide working pH range. Strongly acidic media can be employed because the electrode is 18,000 times more responsive to K^+ than to H^+. Hence, valinomycin-based electrodes exhibit almost ideal characteristics for use in clinical analysis. A Nernstian response to potassium ion activities, with a slope of 59 mV/pK$^+$, is commonly observed from 10^{-6} to 0.1 M. Valinomycin-based potassium electrodes also possess excellent long-term stability. In a comparative study, Griffin and Christian[17] illustrated that aged (10 years) electrodes yielded selectivity data and slopes similar to those of new ones (although response times

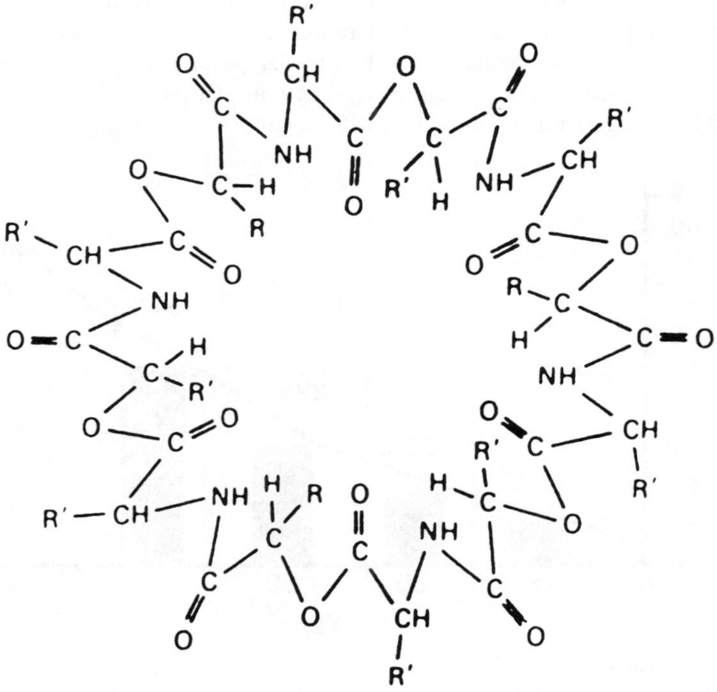

FIGURE 2-5
Valinomycin.

were longer for aged electrodes). As a result of these characteristics, valinomycin-based electrodes are now routinely used for the determination of potassium in whole blood, plasma, serum, and diluted urine. In most cases, plasticized polyvinyl chloride (PVC) membranes are employed. The only problem has been observed during assays of undiluted urine, where a partition of an anionic component of the urine into the PVC-valinomycin membrane occurs.[18] Such interference can be eliminated by using silicone rubber (instead of PVC) as the principal membrane material.[19] Valinomycin-based potassium sensors have been used successfully for various in-vivo or bedside applications, including the measurement of potassium concentration in whole blood during open-heart surgery. Such applications are described in Chapter 5.

Efforts have been made by various groups to replace valinomycin by synthetic potassium carriers.[20, 21] In particular, electrodes containing various bis(crown) ethers exhibit potassium selectivities approaching those of valinomycin-based probes. However, because of their relative poor lipophibility, their overall analytical performance is still not satisfactory.

2-2.2.2 SODIUM MEASUREMENTS

The development of the glass sodium electrode has permitted routine measurements of sodium in whole blood, plasma, serum, and diluted urine. Numerous glass sodium electrodes are available commercially. Most of these are based on glass membranes with compositions of 11% Na_2O– 18% Al_2O_3– 71% SiO_2 or 11% Li_2O– 18% Al_2O_3– 71% SiO_2, which are highly sodium-selective with respect to other alkali metal ions. While the response is often dependent on the conditions of the measurement, the Nernstian range is usually extended from 10^{-6} to 0.1 M. Some lithia-based electrodes will respond according to the Nernst equation down to 5×10^{-9} M. The slope of the calibration plot approaches the theoretical value as the electrode ages (after about 2 months); a new electrode will often yield a slope of 55 mV/decade. Hydrogen ions strongly affect the response, thus limiting the lower pH range in which glass sodium electrodes may be used without error. How low this range is depends on the selectivity coefficient (K_{NaH}) of the particular electrode. Physiological concentrations of cations such as potassium, calcium, or magnesium have a negligible effect on the determination of sodium in blood.[22]

Glass sodium electrodes are now routinely used in commercial electrolyte analyzers. The values obtained for sodium measurements on undiluted systems (i.e., direct potentiometry) are not expected to be the same as those obtained by diluted methods. The direct potentiometric approach measures the activity of sodium only in the water phase, and thus is expected to yield sodium values about 7–9% higher than those obtained using flame photometry. In practice, however, flame photometric values are 96 to 98% of direct potentiometric values. The discrepancy between the values obtained by the two techniques has been the subject of various studies[5, 23, 24] and a large number of letters.[25–27] In addition to considerations of the effective volume occupied by proteins and lipids, binding of sodium ion to bicarbonate may also account for the decreased sodium values obtained in diluted samples.[28] An excellent review of this topic was presented by Levy.[29]

Liquid membrane electrodes based on various carriers are also well suited for sodium measurements in body fluids. These relatively new sensors satisfy the required selectivity for clinical applications and alleviate problems, such as high resistance or protein deposition, which characterize glass membrane electrodes. In addition, liquid membrane electrodes are more suitable for the preparation of sodium microelectrodes.[30] Electrodes containing the carriers ETH 157,[31] ETH 227,[31] and a bis(12-crown-4) compound[32] have been useful for measurements of sodium in blood and urine. The latter exhibits excellent selectivity for sodium over potassium ($K_{NaK} = 0.009$).

2-2.3   Lithium Electrodes

Lithium salts are often used in the treatment of psychiatric patients diagnosed as manic depressive. Because of the therapeutic and toxic effects of lithium, it is important to maintain the lithium blood level over a narrow concentration range (usually between 0.5 to 1.5 mM in blood serum). Reliable monitoring of lithium concentration in serum is thus very important. The major problem with the use of potentiometric electrodes for this purpose is the high sodium content, which is 100-fold greater than the therapeutic lithium level. Lithium ion-selective electrodes, based on different neutral carriers, e.g., macrocyclic crown ethers, acyclic dioxadiamides, or acyclic polyethers, have been proposed to overcome the sodium interference. Detailed investigations of such lithium probes have been performed mainly in the laboratories of Christian[33-36] and Simon.[37, 38] Particularly promising are new lipophilic diamide and 14-crown-4 ether compounds that exhibit relatively high (>200:1) Li-Na selectivity. Considering the high level of sodium, such selectivity is still far from the value desired for a contribution of less than 1% by the activity of the interfering sodium ion. Hence, while these new electrodes can sense lithium in the presence of a large excess of sodium, very careful calibration is required because of small changes in potential over the clinically interesting lithium range (Figure 2-6). Nevertheless, recent studies[35, 36, 38, 39] have indicated the feasibility of potentiometric measurements of lithium in clinically relevant blood samples. The reliability of such measurements can be greatly improved by correction for the sodium response, or through the use of a dialysis membrane in a flow system[39]; the latter will be described in Chapter 4. Simon and coworkers[38] described recently a procedure for measuring the lithium-sodium concentration ratio in blood serum, using a flow cell assembly with lithium- and sodium-selective electrodes. Such a scheme provides useful clinical information because patients with low sodium levels seem to need less lithium to produce the same therapeutic effect.

2-2.4   Calcium Electrodes

Calcium ion activity in various body fluids is known to affect many physiological processes. The calcium ion-selective electrode has proved to be a valuable tool for

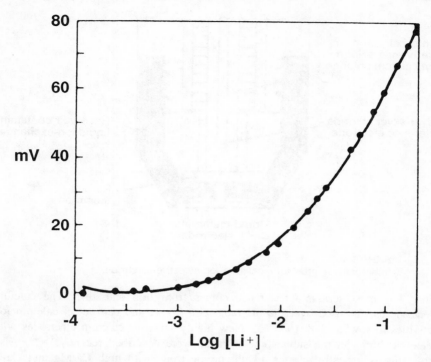

FIGURE 2-6
Response of a lithium-selective electrode in a 140-mM NaCl
solution. (Reproduced with permission.[33])

the determination of calcium in clinical samples. This electrode is a liquid
membrane–type electrode; it uses a liquid cation-exchanger, a solution of calcium
didecyl phosphate in di-n-octylphenyl phosphonate. The ion-exchanger is held in a
porous, plastic filter membrane that separates the test solution from the inner
compartment of the electrode (Figure 2-7). This compartment contains an aqueous
standard solution of calcium chloride into which dips a silver–silver chloride
reference electrode. Upon immersion of the electrode in the sample solution, an
equilibrium is rapidly established at each interface:

$$[(RO)_2POO]_2Ca \rightleftharpoons 2(RO)_2POO^- + Ca^{+2} \qquad (2\text{-}6)$$

The resulting cell potential follows the Nernst equation:

$$E_{cell} = K + (0.059/2)\log[Ca^{+2}] = K - (0.059/2)pCa \qquad (2\text{-}7)$$

The theoretical basis for the calcium electrode was developed by Ruzicka et al.[40]

The calcium electrode is about 200 and 3000 times more selective for calcium
than for magnesium and sodium, respectively. The high selectivity toward calcium
ions is attributed to the great chemical affinity of the organic phosphate for calcium

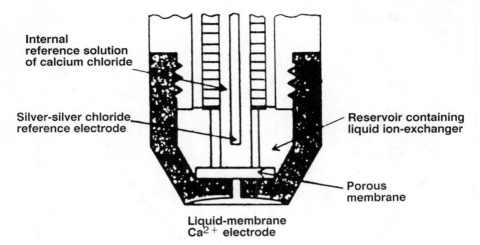

Internal reference solution of calcium chloride

Silver-silver chloride reference electrode

Reservoir containing liquid ion-exchanger

Porous membrane

Liquid-membrane Ca^{2+} electrode

FIGURE 2-7
Schematic diagram of a calcium ion-selective electrode.

ions. The most serious interference comes from iron and zinc. The calcium electrode can be used over the pH range 5.5 to 11, and can detect calcium ion activities as low as 5×10^{-7} M. (New highly sensitive calcium electrodes with detection limits in the subnanomolar range were described recently.[41, 42]) Also promising is a neutral carrier membrane electrode with high Ca-Mg and Ca-H selectivities.[43] The electrode is based on bicyclic polyether amide derivatives, and exhibits rapid response in blood serum.

Unlike monovalent cations with negligible binding and complexing, divalent cations such as calcium exhibit significant protein binding (with more than 50% of the serum calcium being complexed). The physiological importance of free calcium is well established, following the brilliant work of McLean and Hastings.[44] Potentiometric measurements can thus produce information of great clinical significance, which is not readily available otherwise. An important question, however, is whether the calcium-selective electrode yields accurate ionized calcium measurements, as is often claimed. An exact answer is not clear, because of the problem of calcium activity standards (particularly at low levels). The electrode appears to yield an activity value that is operationally defined by the measuring system. Nevertheless, such an index of calcium activity is of much clinical relevance compared with total calcium measurements. Second-generation calcium analyzers, in particular, offer very reliable measurements of ionized calcium in serum, as evidenced from clinical results for a large number of patients' sera.[45] The subject of interpreting ionized calcium measurements has been reviewed.[46, 47]

Potentiometric measurements of calcium in blood, serum, and plasma were evaluated recently.[48] Improved reliability of whole blood analysis was obtained upon collection, handling, and storage under anaerobic conditions and at 0–4°C. Serum samples can be frozen and stored for as long as 6 months without affecting the determination of ionized calcium.[49] Calcium potentiometric measurements have been shown to be affected by anionic surfactants.[50] Such interference can be reduced by changing the solvent mediator in the membrane. The lifetime of calcium probes is usually limited by a gradual loss of the ion-exchanger and cosolvent. Improved stability may be achieved by grafting the cosolvent and the ion-exchanger to the backbone of the polymer matrix.[51] Overall, calcium electrodes exhibit very good precision in assays of serum or whole blood.[45, 52] Measured values may differ when different analyzers are employed,[45, 53, 54] due to differences in the construction of the electrode, the manifold, the salt-bridge solution, or the sample treatment.[55, 56] Various reviews discuss the design and operation of calcium electrodes.[57, 58] Miniature calcium electrodes for in-vivo monitoring are described in Chapter 5.

2-2.5 Halide Ion Electrodes

2-2.5.1 FLUORIDE MEASUREMENTS

The lanthanum fluoride electrode, shown in Figure 2-8, is the most important example of a solid-state electrode. The active membrane of this electrode is a single crystal of lanthanum fluoride doped with europium(II) to lower the electrical resistance and facilitate ionic charge transport. The filling solution contains $0.1\ M$ sodium fluoride and $0.1\ M$ sodium chloride. The electrode exhibits at least a

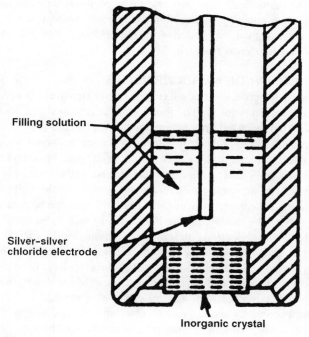

Filling solution

Silver–silver chloride electrode

Inorganic crystal

FIGURE 2-8
Schematic diagram of a fluoride ion-selective electrode.

1000-to-1 preference for fluoride over chloride, bromide, iodide, or bicarbonate ions. The preferential response is linked to mobility within the crystal which is favored by low charge and small ion size. Accordingly, the only significant interferent is a hydroxide ion. Hence, the electrode is limited in use over the pH range of 0 to 8.5 (although when fluoride is in acidic media, it is converted to hydrofluoric acid to which the electrode is insensitive). The cell potential is related to the fluoride ion activity via

$$E = K - 0.059 \log a_{F^-} = K + 0.059 \, pF \tag{2-8}$$

Such a theoretical response is obtained in the range from 10^{-6} to 1 M.

There are a number of clinical applications of the fluoride-selective electrode. While fluoride is not found under normal physiological conditions, various assays of fluoride in individuals who have been exposed to exogenous sources of fluoride have been described. For example, the fluoride-selective electrode was applied for analysis of urine and bone tissues following industrial exposure to fluorine.[59] Similarly, workers at nuclear sites can be surveyed for exposure to fluoride-containing compounds by monitoring their urine.[60] Procedures for the potentiometric measurements of fluoride in whole blood, serum, and plasma were reported.[61, 62] Such procedures allow measurement of nonvolatile fluoride in pediatric patients undergoing methoxyflurane anesthesia.[63] Protein-bound fluorine in blood serum can be estimated following sample pretreatment, e.g., dry ashing or oxygen bomb digestion.[61] An automated standard additions method was developed for the potentiometric determination of free fluoride in plasma and urine.[63] The fluoride ion-selective electrode can be used also for the assessment of patients undergoing renal dialysis and for dental research.

2-2.5.2 CHLORIDE MEASUREMENTS

The chloride-selective electrode has been used extensively for chloride testing in sweat in conjunction with the childhood lung disease cystic fibrosis.[64] Such an electrode is commonly based on a heterogeneous membrane prepared by polymerizing monomeric silicone rubber in the presence of an equal weight of silver chloride particles. A 0.5-mm-thick disk of this heterogeneous membrane is sealed to the bottom of a glass tube; potassium chloride and a silver wire are placed in the tube. The sensitivity of such an electrode is limited by the solubility of silver chloride. Chloride concentrations from 5×10^{-5} to 1.0 M can be measured directly. Oxidizing agents can be used to remove interferences such as bromide, cyanide, and sulfide. A flat-headed electrode design is commonly used for the rapid sweat analysis on the skin (see Chapter 5). Other solid-state selective electrodes for chloride, e.g., based on calomel or silver–silver chloride, have been described.[65, 66] The silver–silver chloride sensor can be used for chloride determination in blood samples, although coverage with a permselective membrane is recommended to minimize interferences by serum proteins that can bind silver ions.[66]

A host of liquid membrane electrodes selective to chloride ions, based on ion-exchangers or neutral carriers (particularly quaternary ammonium compounds), have also been developed over the last decade. The clinical utility of most of these sensors is limited by severe bicarbonate and salicylate interferences. Some of the recently developed liquid membrane probes have reduced interferences from these compounds and allow direct measurements in body fluids.[67, 68]

2-2.5.3 IODIDE MEASUREMENTS

Various solid-state electrodes for iodide based on pressed homogeneous silver iodide–silver sulfide membranes or heterogeneous membranes containing silver iodide in silicone rubber are available. Because of the low solubility of silver iodide, the iodide ion can be measured down to the $10^{-7}M$ level. The homogeneous membrane electrode offers faster response and better stability compared with heterogeneous electrodes. Such solid-state electrodes were applied to the determination of iodide in urine[69, 70] or serum[71]; chloride ions were shown to interfere in urine assays. Another interesting clinical application of iodide electrodes is the determination of the iodine content of thyroid glands.[72, 73]

2-2.6 Miscellaneous Electrodes

Potentiometric electrodes selective to many other ions have been developed. Some of these probes exhibit great promise for clinical analysis.

Behm et al.[74] developed a cyclodecapeptide-based liquid membrane electrode for magnesium with adequate rejection of potassium, lithium, and hydrogen ions, but insufficient Mg-Ca and Mg-Na selectivities. Such an electrode might be useful for certain clinical applications by applying chemometric methods involving simultaneous measurement of the interfering calcium and sodium ions. N,N'-diheptyl-N,N'-dimethylsuccinic acid diamide is another neutral carrier used for potentiometric measurements of magnesium.[75] This lipophilic ligand offers Mg-Ca, Mg-K, and Mg-Na selectivities sufficient for studies of intracellular magnesium activity.

Montalvo et al.[76] developed a redox-type phosphate electrode based on a chemically treated iron wire. The electrode was used for direct measurements of phosphate in serum samples (in the presence of constant oxygen tension). However, such a probe is strongly affected by the redox environment of the sample. Many other attempts to prepare conventional ion-selective electrodes for phosphate have not been successful; recently introduced anion-binding macrocyclic ligands appear to be promising for this task.

Lead measurements can be performed with a solid-state membrane electrode, containing the sulfides of lead and silver. The electrode can be used at concentrations of 10^{-6} and 10^{-1} M lead ion, in the absence of silver and mercury ions. The sensitivity of this electrode is not sufficient for direct measurement of lead in body fluids. The electrode can be used for indirect measurement of anions such as sulfate, phosphate, and oxalate.

The Technicon RA-1000 analyzer incorporates a polymer membrane electrode that is selective for carbonate ions.[77] Blood samples are adjusted to pH 8.6 to liberate the carbonate ions from the bicarbonate and carbon dioxide in the blood. A selective and stable bicarbonate electrode based on planar thin-layer membrane technology was recently described by Oesch et al.[78] Short response times (about 30 s) and a detection limit of 0.2 mM were reported. Other liquid membrane electrodes that are selective for carbonate and bicarbonate ions have been developed[79-82] but have not been applied to clinical analysis.

Electrodes selective to other ions, such as copper, cadmium, silver, nitrate, bromide, cyanide, and perchlorate, are available but have only rarely found a clinical application.

2-2.7   Gas-Sensing Electrodes

The accurate and rapid detection of gases, such as ammonia and carbon dioxide, plays a vital role in the diagnosis of numerous physiological disorders. Gas-sensing electrodes are highly selective devices for measuring dissolved gases. They are reliable and simple, exhibit excellent selectivity, but tend to have relatively slow response times (particularly as the limit of detection is approached). Analyzers based upon such electrodes have been around for almost three decades. Hence, the technology of potentiometric blood gas analysis is clearly a mature one.

Gas-sensing electrodes incorporate a conventional ion-selective electrode surrounded by an electrolyte solution and enclosed by a gas-permeable membrane. The gas (of interest) in the sample solution diffuses through the membrane and comes to equilibrium with the internal electrolyte solution. In the internal compartment, between the membrane and the ion-selective electrode, the gas undergoes a chemical reaction, consuming or forming an ion to be detected by the ion-selective electrode. (Protonation equilibria in conjunction with a pH electrode are most common.) Since the activity of this ion is proportional to the amount of gas dissolved in the sample, the electrode response is directly related to the concentration of the gas in the sample. The response is usually linear over a range of typically four orders of magnitude; the upper limit is determined by the concentration of the inner electrolyte solution. The permeable membrane is the key to the electrode's gas selectivity. Two types of polymeric material, microporous and homogeneous, are used to form the gas-permeable membrane. Such hydrophobic membranes are impermeable to water or ions. Hence, gas-sensing probes exhibit excellent selectivity, compared with many ion-selective electrodes. Besides the membrane, the response characteristics are often affected by the composition of the internal solution and the variables of geometry.[83]

Nonmembrane gas-sensing probes, known as *air-gap* electrodes, have also been employed. In principle, such probes do not differ significantly from membrane-based sensors. The air-gap electrode uses an actual air gap (several millimeters thick) across which gas diffusion takes place toward the inner electrolyte. The latter is usually wicked to the surface of the glass-sensing element with the aid of a wetting agent or is supported by a special sponge. The relative merits of membrane and nonmembrane gas-sensing probes have been discussed.[84]

Gas sensors for carbon dioxide and ammonia have been very successful in the clinical arena, as described in the following sections. Also described below are oxygen probes, although they are based upon amperometric sensing. Other clinical applications of potentiometric gas probes, including their utility as sensing elements in enzyme electrodes or as in-vivo devices, are described in Chapters 3 and 5, respectively.

2-2.7.1 CARBON DIOXIDE SENSORS

Carbon dioxide devices were originally developed by Severinghaus and Bradley[85] to measure the partial pressure of carbon dioxide in blood. This electrode, still in use today (in various automated systems for blood gas analysis), consists of an ordinary glass pH electrode covered by a carbon dioxide membrane, usually silicone, with an electrolyte (sodium bicarbonate–sodium chloride) solution entrapped between (Figure 2-9). When carbon dioxide from the outer sample diffuses through the semipermeable membrane, it lowers the pH of the inner solution:

$$CO_2 + H_2O \rightleftharpoons HCO_3^- {}_{+H^+} \tag{2-9}$$

Such changes in the pH are sensed by the inner glass electrode. The overall cell potential is thus determined by the carbon dioxide concentration in the sample:

$$E = K + (RT/F)\, ln[CO_2] \tag{2-10}$$

Such a Nernstian response of 59-mV/decade change in concentration is commonly observed (at 25°C). Relation to the partial pressure carbon dioxide is accomplished by the use of Henry's law.

The selectivity characteristics of the carbon dioxide electrode have been the subject of various studies.[87, 88] Interferences, e.g., benzoic acid, have been reported using certain carbon dioxide analyzers.[89] Kobos et al.[87] illustrated that each type of gas-permeable membrane has a pronounced effect on the observed selectivity, with the commonly used silicone rubber resulting in large response to nonvolatile organic

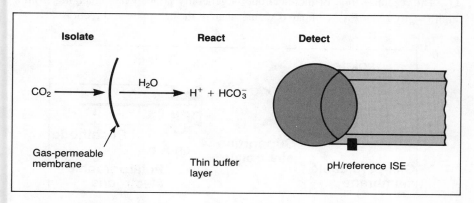

FIGURE 2-9
Potentiometric carbon dioxide electrode. (Reproduced with permission.[86])

acids. Homogeneous Teflon membranes result in improved selectivity but at the cost of slow response times. Other factors affecting the response time, including solution temperature, carbonic acid–carbon dioxide equilibria, and the internal filling solution sodium bicarbonate concentration, were investigated.[90] Simon and coworkers[91] explored the behavior of carbon dioxide electrodes in the presence of interfering ions and illustrated that the selectivity coefficients are time-dependent quantities. Several other configurations of potentiometric carbon dioxide probes have been developed and shown to offer certain advantages.[92, 93]

2-2.7.2 OXYGEN ELECTRODES

Considerable attention has been directed toward the measurement of the oxygen partial pressure (P_{O_2}) in blood. Membrane-covered oxygen probes based on Clark's design[94] have found acceptance for this and other clinical applications. The sensor is based on a pair of electrodes immersed in an electrolyte solution and separated from the test solution by a gas-permeable hydrophobic membrane (Figure 2-10). The membrane is usually made of Teflon, silicone, or polyethylene, while the electrolyte is a solution of potassium chloride and buffer. Oxygen diffuses through the membrane and is reduced at the surface of the sensing electrode. The resulting electrolytic current is proportional to the rate of diffusion of oxygen to the cathode, and hence to the partial pressure of oxygen in the sample. The actual potential applied at the cathode (with respect to the anode/reference electrode) depends on the particular design. Cathodes made of platinum, gold, or silver are commonly incorporated in different commercial probes. The applied potential usually maintains the cathode on the diffusion-limited plateau region for the oxygen reduction process. A lowering of the potential is recommended in the presence of possible reducible interferences, e.g., halogenated hydrocarbons used as anesthetics, including the common one — halothane.[95] The electrode is usually calibrated by exposure to samples with known oxygen content, e.g., with air assumed 20.93% O_2. The response time of the electrode is generally larger when changing from a high P_{O_2} to a low P_{O_2}, compared with a change in the opposite direction.

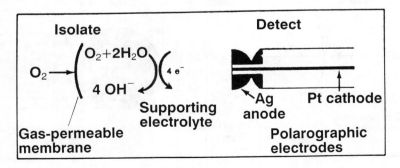

FIGURE 2-10
Membrane-covered oxygen electrode — Clark's design.
(Reproduced with permission.[86])

Commercial blood gas analyzers commonly couple P_{O_2} measurements with P_{CO_2} and pH measurements. Complete analysis of $50-150$-μl blood samples can be accomplished in about 3 min. While these analyzers are based on early electrode designs (described before), they incorporate microprocessor-based electronics for improved automation, data acquisition, and signal processing. The main challenges in blood gas measurements based on these analyzers is the preservation of the sample's integrity (loss of gasses or contamination) and the control of its temperature (as the current response exhibits strong temperature dependence). Other minor problems include flow artifacts and current drifts.

In-vivo sensing of oxygen (invasive or "through the skin") is described in Chapter 5. The coupling of enzymes with the Clark electrode, which provides the basis for various biosensors, is described in Chapter 3.

2-2.7.3 AMMONIA PROBES

The determination of ammonia in blood is an important diagnostic test for several disease states (e.g., Reye's syndrome, hepatic coma). Commercially available ammonia-selective gas sensors use a pH glass membrane as an internal sensing element, in conjunction with an ammonium chloride inner electrolyte solution. The glass electrode detects a decrease in the activity of hydrogen ions from the hydroxyl ions produced in the internal solution:

$$NH_3 + H_2O \rightleftarrows NH_4^+ + OH^- \tag{2-11}$$

Such an electrode requires operation at a basic solution (pH > 10.5), so that all ammonia in the sample is present as free dissolved gas. Sensitivity usually extends from 10^{-6} to 1 M. The electrode has been widely used for the determination of ammonia in blood using both static and automated flow systems.[96-98] Improvements in the dynamic performance of ammonia gas sensors have been reported.[99] The selectivity of the ammonia gas electrode was examined.[100] Interferences due to volatile amines, as well as various amino acids (undergoing hydrolysis, which liberates ammonia), have been reported. The anticoagulant used for collecting the blood samples may also affect the data, via its influence on the distribution of ammonia between plasma and the red blood cells.[101] Existing probes require large ($3-6$-ml) samples, which make the method impractical for use in newborns or small children.

Meyerhoff[102] developed a polymer membrane–based ammonia gas electrode that offers improved detection limits (over existing commercial devices) and allows measurements under milder conditions (pH 8.5), i.e., closer to the physiological pH. This probe is based on the measurement of ammonium ions formed inside the film of buffer sandwiched between a polymer ammonium-sensitive membrane and the gas-permeable membrane (Figure 2-11). The use of mild buffer conditions reduces interferences due to hydrolysis of amino acids. New disposable ammonia sensors based upon this electrode were developed.[103] Such probes require less than 250- μl samples and thus are very attractive for assays performed in the physician's office or in a small laboratory. Another micromethod for the determination of

[a]

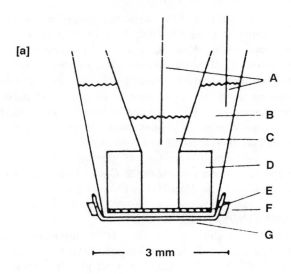

[b]

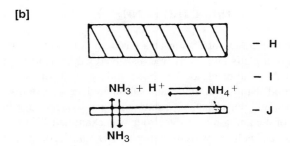

FIGURE 2-11
(a) Ammonia sensor: (A) Ag-AgC*l* electrodes, (B) internal buffer,
(C) 0.01 *M* NH$_4$C*l*, (D) tubing, (E) PVC membrane, (F) O-ring, (G)
gas-permeable membrane. (b) Processes taking place at the tip:
(H) polymer-nonactin membrane, (I) buffer film, (J) gas-permeable
membrane. (Reproduced with permission.[102])

plasma ammonia permits accurate measurements using 200-μ*l* samples.[104] The
air-gap electrode configuration can also be clinically useful, as was illustrated in
ammonia measurements in blood.[105]

2-2.8   Coated Solid-State Devices

The need to make potentiometric measurements in very small volumes, or in-vivo,
has prompted the development of coated-wire electrodes (CWEs) and ion-selective

field effect transistors (ISFETs). The significant size reduction of these solid-state devices is achieved by removal of the internal reference solution.

Coated-wire electrodes, introduced by Freiser in the mid-1970s, are prepared by coating an appropriate polymeric film directly onto a conductor (Figure 2-12). The ion-responsive membrane is commonly based on polyvinyl chloride, while the conductor can be metallic (Pt, Ag, Cu) or graphite-based of any conventional shape, e.g., wire, disk. The conductor is usually dipped in a solution of PVC and the active substance, and the resulting film is allowed to air-dry. Other polymers and modified polymers, including polyacrylic acid and modified polyvinylbenzyl chloride, can also be useful for various applications. In addition to the miniaturization capability, CWEs are extremely simple, inexpensive, and easy to prepare and function well over the 10^{-5}- to $0.1M$ concentration range. The exact mechanism of the CWE

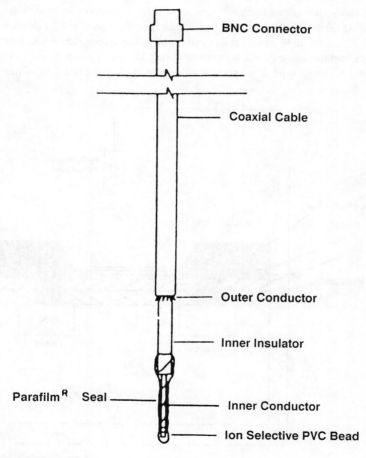

FIGURE 2-12
Coated-wire ion-selective electrode. (Reproduced with permission.[106])

behavior continues to be a mystery, in view of the lack of internal reference components. CWEs may suffer from reproducibility and long-term stability (drifting potential) problems. Nevertheless, such devices have been found useful for various important applications, provided that the electrodes are calibrated periodically. The determination of basic drugs, e.g., cocaine, methodane,[107] amino acids,[108] potassium and sodium,[109] lithium,[35] and calcium,[110] represents some of the clinically useful applications of CWEs. New concepts for preparing CWEs appear to improve their analytical performance, particularly with respect to stability and reproducibility. The principles and applications of CWEs have been reviewed.[111]

Ion-selective field effect transistors offer an alternative approach to miniaturization of potentiometric probes. The development of ISFETs is considered as a logical extension of CWEs. The construction of ISFETs is based on the technology used to fabricate microelectronic chips. ISFET incorporates the ion-sensing membrane directly on the gate area of a field effect transistor (FET) (Figure 2-13). The FET is a solid-state device that exhibits high-input impedance and low-output impedance and therefore is capable of monitoring charge buildup on the ion-sensing membrane. As the charge density on this membrane changes because of interaction

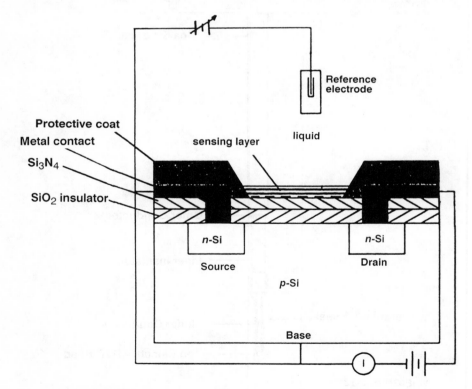

FIGURE 2-13
An ion-selective field effect transistor.

with the ions in solution, a drain current is flowing between the source and drain of the transistor. The increased voltage needed to bring the current back to its initial value represents the response. (This is commonly accomplished by placing the ISFET in a feedback loop.) From the standpoint of change in drain current as a result of change in activity of the ion of interest, the ISFET response is governed by the same Nernstian relationship (and the selectivity limitation) that characterizes conventional ion-selective electrodes.

Such sensors that utilize solid-state electronics have significant advantages. The actual sensing area is very small. Hence, a single miniaturized solid-state chip could contain multiple gates and be used to sense several ions simultaneously. Other advantages include the in-situ impedance transformation and the ability for temperature and noise compensation. While the concept of ISFET is thus very exciting and intriguing, problems with stability and encapsulation still need to be solved before such devices reach the truly practical stage. One problem is the detachment of PVC-type ion-sensing membranes from the gates of FETs. This problem can be minimized by suspending a polyimide mesh over the gate[112]; the polymer film thus becomes anchored in place by the mesh.

The coating on the gate is the key to the analytical chemistry that the ISFET can perform. ISFETs based on various ion-responsive layers have been developed. Among these are a sodium ISFET based on the synthetic sodium carrier ETH 227, an ammonium ISFET utilizing monactin-nonactin,[113] and a chloride ISFET prepared by laying a membrane of methyltridodecylammonium chloride over a silicon nitride gate.[114] ISFETs that are not covered with an ion-responsive membrane can be used directly as pH sensors. The silicon nitride coat on the transistor is itself sensitive to hydrogen ions (by its own surface properties), and develops phase boundary potentials proportional to the logarithm of the hydrogen ion activity in the contacting solution. The feasibility of sensing several ions was illustrated using a quadruple-function ISFET probe that simultaneously monitored potassium, sodium, calcium, and pH in whole blood samples.[115] ISFETs can be combined with various biological agents, e.g., enzymes, antigens, to form effective biosensors (see Chapter 3). The theory and mode of operation of ISFETs have been reviewed.[116]

2-2.9   Multilayer Films for Potentiometric Measurements

Multilayer film technology, developed by Eastman Kodak,[117] offers a new concept for potentiometric systems for clinical analysis. Measurements are performed on the surfaces of disposable ion-selective electrode slides, responsive to potassium, sodium, chloride, and carbon dioxide. The slides are prepared by coating appropriate polymeric membranes over internal reference gel–electrode coatings on conductive substrates. For example, a schematic diagram of a cell used for potassium determination in serum samples is shown in Figure 2-14. Here, two identical valinomycin-based thin-film electrodes are coupled by a paper salt bridge—essentially a miniature electrochemical cell in a plastic mount. One drop (10 μl) of blood serum is applied to one electrode, and an equal volume of standard

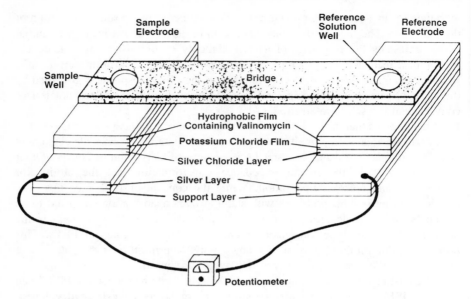

FIGURE 2-14
Schematic of an Eastman Kodak Ektachem slide for
potentiometric measurements of potassium. (Reproduced with
permission.[118])

potassium solution to the other. The two fluids form a liquid junction by capillary
flow through a small strip of ion-free paper, which replaces the traditional salt
bridge. The paper strip is coated with polyethylene to confine the fluid flow and
inhibit evaporation. A potentiometer connected to the electrodes measures the
resulting potential difference. No preconditioning of the dry electrode is required,
and a stable potential is reached within 3 min after placing the liquid test sample.
This information is transmitted to the microprocessor, which calculates and displays
the potassium concentration of the patient's serum.

The performance of these dry-reagent slides has been shown to be comparable
to that of conventional ion-selective electrodes. Because each sample is done on a
fresh electrode, problems of electrode fouling or poisoning are eliminated. The slide
contains all the necessary reagents (except 10 μl of the reference fluid), and thus no
reagents must be stored or mixed. Rapid analysis can be obtained with freshly
drawn small blood samples. The small sample volumes make this approach
attractive for neonatal and geriatric patients.

Analyzers based on the dry-reagent technology (available from Eastman
Kodak, Miles Labs, or Sentech Corp.) are small and require little technical skill to
operate. The Sentech analyzer (Model Chempro-1000) can measure four parameters
(K, Na, pH, Ca) within 60 s using 150-μl samples.[119] Such analyzers are ideally
suited for use in physicians' offices and in critical-care settings and for home health
care.

2-2.10 Conclusion

Ion-selective electrodes have very exciting possibilities for clinical analysis. With the help of these devices, various ions and gases can be easily monitored in blood, urine, sweat, etc. While detection of cationic electrolytes in body fluids is now routine, analogous measurements of anions (e.g., HCO_3^-, PO_4^{-3}, etc.) remain a difficult task. It won't be too long, however, before we can measure all the key electrolytes reliably by ion-selective electrodes. The availability of reliable electrode-based electrolyte and gas analyzers has resulted in rapid acceptance of potentiometry by clinicians. However, more discussion among clinicians, scientists, and instrument manufacturers is desired to establish the working basis for the standardization of measurement techniques and nomenclature. New technologies based on thin films and semiconductor chips are extremely useful to meet future requirements, trends, and challenges. For more information on the utility of ion-selective electrodes for clinical analysis, the reader is referred to References 4 and 120.

REFERENCES

1. Buck, R. P., *CRC Crit. Rev. Anal. Chem., 5*, 323 (1976).
2. Covington, A. K., *CRC Crit. Rev. Anal. Chem., 3*, 355 (1974).
3. Ammann, D., Morf, W. Z.., Anker, P., Meier, P. C., Pret, E., Simon, W., *Ion Sel. Elec. Rev., 5*, 3 (1983).
4. Oesch, U., Ammann, D., Simon, W., *Clin. Chem., 32*, 1448 (1986).
5. Koch, D., Graves, S., Ladenson, J. H., *Clin. Chem., 28*, 1931 (1982).
6. Meyerhoff, M. E., Fraticelli, Y. M., *Anal. Chem., 54*, 27R (1982).
7. Arnold, M. A., Meyerhoff, M. E., *Anal. Chem., 56*, 20R (1984).
8. Arnold, M. A., Solsky, R. L., *Anal. Chem., 58*, 84R (1986).
9. Moody, G. J., Thomas, J. D. R., *Ion-Sel. Elec. Rev., 6*, 209 (1984).
10. Koryta, J., *Anal. Chim. Acta, 139*, 1 (1982).
11. Eisenman, G. (ed.), *Glass Electrodes for Hydrogen and Other Cations*, Marcel Dekker, New York, 1976.
12. Anker, P., Ammann, D., Simon, W., *Mikrochim. Acta, I*, 237 (1983).
13. Christian, G. D., *Analytical Chemistry*, 4th ed., Wiley, New York, 1986, pp. 122, 303.
14. Maas, A. H. J., Weisberg, H. F., Ziljstra, W. G., Durst, R. A., Siggard-Andersen, O., *J. Clin. Chem. Clin. Biochem., 21*, 313 (1983).
15. Savory, J., Bertholf, R. L., Boyd, J. C., Bruns, D. E., Felder, R. A., Lovell, M., Shipe, J. R., Wills, M. R., Czaban, J. D., Coffey, K. F., O'Connell, K. M., *Anal. Chim. Acta, 180*, 99 (1986).
16. Maas, A. H. J., Siggaard-Andersen, O., Weisberg, H. F., Ziljstra, W. G., *Clin. Chem., 31*, 482 (1985).
17. Griffin, J. J., Christian, G. D., *Talanta, 30*, 201 (1983).
18. Jenny, H. B., Riess, C., Ammann, D., Magyar, B., Asper, R., Simon, W., *Mikrochim. Acta, II*, 309 (1980).
19. Anker, P., Jenny, H. B., Wuthier, U., Asper, R., Ammann, D., Simon, W., *Clin. Chem., 29*, 1447 (1983).
20. Kimura, K., Maeda, T., Tamura, H., Shono, T., *J. Electroanal. Chem., 95*, 91 (1979).

21. Linder, E., Toth, K., Horvath, M., *Z. Anal. Chem., 322,* 157 (1985).
22. Bijster, P., Vink, K. L. J., *Clin. Chem., 30,* 865 (1984).
23. Czaban, J. D., Cormier, A. D., Legg, K. D., *Clin. Chem., 28,* 1936 (1982).
24. Kissel, T. R., Sandifer, J. R., Zumbulyadis, N., *Clin. Chem., 28,* 449 (1982).
25. Coleman, R. L., *Clin. Chem., 25,* 1865 (1979).
26. Czaban, J. D., Cormier, A. D., *Clin. Chem., 26,* 1921 (1980).
27. Shyr, C., Young, C. C., *Clin. Chem., 26,* 1517 (1982).
28. Coleman, R. L., Young, C. C., *Clin. Chem., 28,* 1705 (1982).
29. Levy, G. B., *Clin. Chem., 27,* 1435 (1981).
30. Steiner, R. A., Oehme, M., Ammann, D., Simon, W., *Anal. Chem., 51,* 351 (1979).
31. Anker, P., Jenny, H. B., Wuthier, U., Asper, R., Ammann, D., Simon, W., *Clin. Chem., 29,* 1508 (1983).
32. Shono, T., Okara, M., Ikeda, I., Kimura, K., *J. Electroanal. Chem., 132,* 99 (1982).
33. Gadzekpo, V. P. Y., Hungerford, J. M., Kadry, A. M., Ibrahim, Y. A., Christian, G. D., *Anal. Chem., 57,* 493 (1985).
34. Gadzekpo, V. P. Y., Hungerford, J. M., Kadry, A. M., Ibrahim, Y. A., Xie, R. Y., Christian, G. D., *Anal. Chem., 58,* 1948 (1986).
35. Xie, R. Y., Christian, G. D., *Analyst, 112,* 61 (1987).
36. Zhou, Z. N., Xie, R. Y., Christian, G. D., *Anal. Lett., 19,* 1747 (1986).
37. Metzger, E., Ammann, D., Asper, R., Simon, W., *Anal. Chem., 58,* 132 (1986).
38. Metzger, E., Dohner, R., Simon, W., Vonderschmitt, D. J., Gautschi, K., *Anal. Chem., 59,* 1600 (1987).
39. Xie, R. Y., Christian, G. D., *Anal. Chem., 58,* 1806 (1986).
40. Ruzicka, J., Hansen, E. H., Tjell, J. C., *Anal. Chim. Acta, 67,* 155 (1973).
41. Morton, R. W., Chung, J. K., Miller, J. L., Charlton, J. P., Fager, R. S., *Anal. Biochem., 157,* 345 (1986).
42. Schefer, V., Ammann, D., Pretsch, E., Oesch, V., Simon, W., *Anal. Chem., 58,* 2282 (1986).
43. Kimura, K., Kumani, K., Kitazawa, S., Shono, T., *Anal. Chem., 56,* 2369 (1984).
44. McLean, F. C., Hastings, A. B., *Am. J. Med. Sci., 189,* 601 (1935).
45. Bowers, G. N., Brassard, C., Sena, S. F., *Clin. Chem., 32,* 1437 (1986).
46. Siggaard-Andersen, O., Thode, J., *Scand. J. Clin. Lab. Invest. Suppl., 43,* 11 (1983); *Chem. Abstr., 99,* 19156W (1983).
47. Ladenson, J. H., Bowers, G. N., *Clin. Chem., 19,* 565 (1973).
48. Wandrup, J., Kvetny, J., *Clin. Chem., 31,* 356 (1985).
49. Plant, S. B., McCarron, D. A., *Clin. Chem., 28,* 1362 (1982).
50. Hulanicki, A., Trajanowicz, M., Augostowska, M., *Z. Anal. Chem., 26,* 115 (1981).
51. Hobby, P. C., Moody, G. J., Thomas, J. D. R., *Analyst, 108,* 581 (1983).
52. Fogh-Andersen, N., Christiansen, T. F., Komarmy, L., Siggaard-Andersen, O., *Clin. Chem., 24,* 1545 (1978).
53. Ohaman, S., Larsson, L., *Clin. Chem., 24,* 2070 (1978).
54. Demetriou, L., *Clin. Chem., 24,* 2071 (1978).
55. Fuchs, C., McIntosh, C., *Clin. Chem., 23,* 610 (1977).
56. Larsson, L., Ohman, S., *Clin. Chem., 26,* 1761 (1980).
57. Robertson, W. G., Marshall, R. W., *CRC Crit. Rev. Clin. Lab. Sci., 11,* 271 (1979).
58. Moody, G. J., Thomas, J. D. R., *Ion-Sel. Elec. Rev. 1,* 3 (1979).
59. Irlweck, K., Czitober, H., Machata, G., *Acta Med. Austriaca, 6,* 99 (1979).
60. Pires, M. A., Abrao, A., *Am. Simp. Bras. Electroquim. Electroanal, 4th. Chem. Abstr., 101,* 103136g (1984).
61. Chiba, K., Tsunoda, K. I., Fuwa, K., *Anal. Chem., 52,* 1582 (1980).
62. Kissa, E., *Clin. Chem., 33,* 253 (1987).
63. Phillips, K. A., Rix, C. J., *Anal. Chim. Acta, 169,* 263 (1985).
64. Moody, G. J., Thomas, J. D. R., *Ion-Sel. Elec. Rev., 2,* 73 (1980).
65. Serikov, Y. A., Komova, V. I., *Zavod. Lab., 49,* 8 (1983).

66. Seshimoto, O., Sakaguchi, S., Takayama, T., Sato, A., Ger. Offen DE (Patent) 3,222,464 (Dec. 30, 1982); *Chem. Abstr., 98,* 85779g (1983).
67. Willis, J. P., Young, C. C., Martin, R., Stearns, P., Pelosi, M., Magnanti, D., *Clin. Chem., 29,* 1193 (1983).
68. Hitachi, Ltd. Jpn. Tokyo Koho JP (Patent) 5757,655 (Dec. 6, 1982); *Chem. Abstr., 99,* 35661n (1983).
69. Copper, G. J. S., Croxson, M. S., *Clin. Chem., 29,* 1320 (1983).
70. Morin, P. P., Caroff, J., Savina, A., Thomas, J., Lahellec, M., Morin, J. P., *Ann. Bil. Clin., 33,* 89 (1975).
71. Carter, R. A., *Proc. Assoc. Clin. Biochem., 5,* 67 (1968).
72. Puttemans, F., Deconinck, F., Jonckheer, M., Vandeputte, M., Massart, D. L., *Clin. Chem., 25,* 1247 (1979).
73. Vajgand, V. J., *Glas. Hem. Drus. Beograd., 49,* 113; *Chem. Abstr., 101,* 104251C (1984).
74. Behm, F., Ammann, D., Simon, W., Brunfeldt, K., Halstrom, J., *Helv. Chim. Acta, 68,* 110 (1985).
75. Lanter, F., Erne, D., Ammann, D., Simon, W., *Anal. Chem., 52,* 2400 (1980).
76. Montalvo, J. G., Truxillo, L. A., Wawro, R. A., Watkins, T. A., Phillips, A., Jenevin, R. M., *Clin. Chem., 28,* 655 (1982).
77. Chapoteau, E., Scott, W. J., *Clin. Chem., 29,* 1187 (1983).
78. Oesch, U., Malinowska, E., Simon, W., *Anal. Chem., 59,* 2131 (1987).
79. Greenberg, J. A., Meyerhoff, M. E., *Anal. Chim. Acta, 141,* 57 (1982).
80. Mostert, I. A., Morf, W. E., Simon, W., *Mikrochim. Acta, 3,* 245 (1984).
81. Grekovich, A. L., Materova, E. A., Garbuzova, N. W., *Z. Anal. Khim., 28,* 1206 (1973).
82. Wise, M. W., U.S. Patent No. 3 723 281 (1973).
83. Ross, J. W., Riseman, J. H., Kruger, J. A., *Pure Appl. Chem., 36,* 473 (1973).
84. Bailey, P. L., Riley, M., *Analyst, 100,* 145 (1975).
85. Severinghaus, J. W., Bradley, A. F., *J. Appl. Physiol., 13,* 515 (1957).
86. Czaban, J. D., *Anal. Chem., 57,* 345A (1985).
87. Kobos, R. K., Parks, S. J., Meyerhoff, M. E., *Anal. Chem., 54,* 1976 (1982).
88. Lopez, M., *Anal. Chem., 56,* 2360 (1984).
89. Steelman, M., *Clin. Chem., 30,* 562 (1984).
90. Jensen, M. A., Rechnitz, G. A., *Anal. Chem., 51,* 1972 (1979).
91. Morf, W. E., Mostret, I. A., Simon, W., *Anal. Chem., 57,* 1122 (1985).
92. Meyerhoff, M. E., Patricelli, Y. M., Greenberg, J. A., Rosen, J., Park, S., Opdycke, W. N., *Clin. Chem., 28,* 1973 (1982).
93. Scarano, E., Naggar, P., Belli, R., *Anal. Lett., 16,* 723 (1983).
94. Clark, L. C., Wolf, R., Granger, D., Taylor, Z., *J. Appl. Physiol., 6,* 89 (1953).
95. Severinghaus, J. W., Wieskopf, R. B., Nishimura, M., Bradley, A. F., *J. Appl. Physiol., 31,* 640 (1971).
96. Park, N. J., Fenton, J. C. B., *J. Clin. Pathol., 26,* 802 (1973).
97. Sanders, G. T. B., Thornton, W., *Clin. Chim. Acta, 46,* 465 (1973).
98. Coleman, R. L., *Clin. Chem., 18,* 867 (1972).
99. Guilbault, G. G., Czarnecki, J. P., Nabi Rahmi, M., *Anal. Chem., 57,* 2110 (1985).
100. Lopez, M. E., Rechnitz, G. A., *Anal. Chem., 54,* 2085 (1982).
101. Davidson, J. S., Jennings, D. B., *Can. J. Physiol. Pharmacol., 58,* 550 (1980).
102. Meyerhoff, M. E., *Anal. Chem., 52,* 1532 (1980).
103. Meyerhoff, M. E., Robin, R. H., *Anal. Chem., 52,* 2383 (1980).
104. Cooke, R. J., Jensen, R. L., *Clin. Chem., 29,* 867 (1983).
105. Ruzicka, J., Hansen, E. A., *Anal. Chim. Acta, 69,* 129 (1974).
106. Martin, C. R., Freiser, H., *J. Chem. Ed., 57,* 512 (1980).
107. Cunningham, L., Freiser, H., *Anal. Chim. Acta, 139,* 97 (1982).
108. James, H., Carmack, G., Freiser, H., *Anal. Chem., 44,* 856 (1972).

109. Tamura, H., Kimura, K., Shono, T., *Anal. Chem., 54*, 1224 (1982).
110. Heidecke, G., Stork, G., Schindler, J. G., von Geulich, M., Schmid, W., Maier, H., Lindt, H. O., Sailer, D., *Z. Anal. Chem., 301*, 406 (1980).
111. Freiser, H., "Coated Wire Ion-Selective Electrodes," in H. Freiser (ed.), *Ion-Selective Electrodes in Analytical Chemistry*, Plenum Press, New York, vol. 2, p. 85.
112. Blackburn, G., Janata, J., *J. Electrochem. Soc., 129*, 2580 (1982).
113. Oesch, V., Caras, S., Janata, J., *Anal. Chem., 53*, 1983 (1981).
114. Zhukova, T. V., *Zavod. Lab., 50*, 18 (1984); *Chem. Abstr., 101*, 221484S (1984).
115. Sibbard, A., Covington, A. K., Carter, R. F., *Clin. Chem., 30*, 135 (1984).
116. Janata, J., Huber, R. J., *Ion-Sel. Elec. Rev., 1*, 31 (1979).
117. *Chem. Eng. News,* 33 (Aug. 11, 1980).
118. Walter, B., *Anal. Chem., 55*, 498A (1983).
119. Harding, R. S., *Clin. Chem., 32*, 1105 (1986).
120. Solsky, R. L., *CRC Crit. Rev. Anal. Chem., 14*, 1 (1985).

C H A P T E R 3

Electrochemical Biosensors

3-1 INTRODUCTION

Biosensors are devices employing biochemical molecular recognition properties as the basis for a selective bioanalysis. The major processes involved in any biosensor system are analyte recognition, signal transduction, and readout. Electrochemical devices have traditionally received a major share of the attention in biosensor development. This is because of the inherent simplicity of obtaining a direct electrical readout and the opportunities for using integrated circuit technology.

Electrode-based biosensors have historically made use of immobilized enzymes, but new strategies using intact microorganisms, biological tissues, antigens or antibodies, and other molecular recognition elements have been shown successful for various clinical purposes. These include the determination of therapeutic drugs, antigens, antibodies, and metabolites. The aim is to produce a signal (potential, current, etc.) proportional to the concentration of the species of interest, utilizing its specific biochemical or biophysical process. Often the product of the biocatalytic process is sensed by the electrochemical transducer (as illustrated in Figure 3-1). Intimate contact between the biosystem and the transducer is achieved by immobilization at the transducer surface, generally by one of several methods (chemical cross-linking, physical entrapment, etc). Depending upon the specific interacting system, different types of clinical information can be achieved. The high affinity of immunological systems can be used to measure trace levels of various metabolites. In contrast, enzymes display lower affinity toward their substrates that can be used in fully reversible sensors. Amperometric and potentiometric transducers are most commonly used in conjunction with electrochemical biosensors. An amperometric biosensor may be more attractive because of its high sensitivity and wide linear range. Matrix effects (upon the amperometric response) can often be minimized by the judicious choice of protective membranes. Potentiometric probes are usually less prone to interferences. In particular, gas-sensing potentiometric probes exhibit superior selectivity (although they suffer from long response and recovery times).

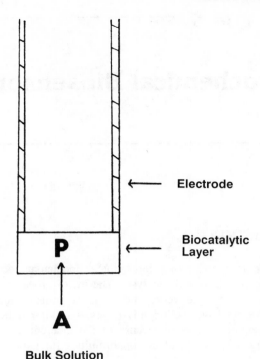

←—— **Electrode**

←—— **Biocatalytic
            Layer**

P

↑

A

Bulk Solution

FIGURE 3-1
Electrode-based biosensor. A—analyte; P—product of biocatalytic
reaction.

Since the introduction of the first enzyme electrode in 1961, over 100 different
types of bioelectrodes have been proposed for measuring numerous clinically
significant compounds. Such sensors may offer many potential advantages that
make them ideally suited for many applications in clinical diagnostics. These
advantages include simple and low-cost instrumentation, fast response times,
minimum sample pretreatment, and high sample throughput. Clinical tests
performed at the physician's office or at the patient's bedside or home can benefit
particularly from these advantages. There are, however, several problems to be
overcome to bring electrochemical biosensors into widespread routine clinical use.
These include the limited working lifetime or dynamic range, the skills required for
construction of biosensors, and the need for frequent calibration. Electrode lifetimes
are typically dictated by the biocatalyst instability, and may range from 1 day to
several months. The above problems will be alleviated with the rapid growth of
biosensor technology. Coupled with other advances (e.g., new biocatalysts or
biomaterials), new and more powerful electrochemical biosensors are expected.

This chapter will attempt to provide an overview of the rapidly developing field
of electrochemical biosensors. In-vivo (implantable) sensors will be discussed in
Chapter 5.

3-2 ENZYME ELECTRODES

3-2.1 Principles

Enzymes are proteins that catalyze chemical reactions in living systems. They act according to the scheme:

$$\text{Substrate} + \text{cofactor} \xrightarrow{\text{enzyme}} \text{products} \qquad (3\text{-}1)$$

Such catalysts are not only efficient, but also often remarkably selective. For example, an enzyme is capable of catalyzing a particular reaction of a particular substrate even though other isomers of that substrate, or similar substrates, may be present. Because of their specificity and sensitivity, enzymes continue to enjoy widespread use as analytical tools in clinical laboratories. An enzyme electrode combines the specificity and affinity of an enzyme for its substrate with the analytical power of electrochemical devices. As a result of such coupling, enzyme electrodes have been shown to be extremely useful for monitoring a variety of substrates of clinical importance.

Enzyme-based electrochemical sensors can be divided into two groups according to their principle of operation: amperometric and potentiometric. Such devices are usually prepared by attaching an immobilized enzyme layer to the electrode layer, which monitors changes occurring as a result of the enzymatic reaction potentiometrically or amperometrically. A schematic representation of a common amperometric device is shown in Figure 3-2. Batch (static) probes,

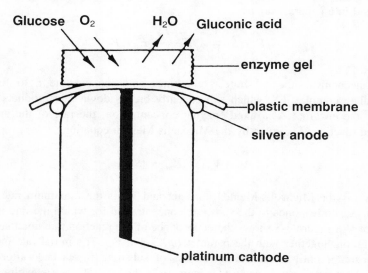

FIGURE 3-2
Schematic diagram of an amperometric glucose electrode.
(Reproduced with permission.)

flow-through reactors, and microelectrodes, based on these configurations, have been devised, and several enzyme-electrochemical systems are available commercially. The choice of the sensing element depends primarily upon the enzymatic system employed. For example, amperometric probes are highly suitable when oxidase enzyme – based systems are employed. Carbon dioxide gas potentiometric electrodes will be the choice when decarboxylase enzymes are used.

The success of the enzyme electrode depends, in part, on the immobilization of the enzyme layer. The technology of immobilized enzymes has experienced phenomenal growth in the recent past. A book[2] and several reviews[3-5] describe various approaches for attaching enzymes to different matrices. Chemical and physical immobilization schemes are commonly employed in conjunction with electrochemical sensors. In chemical immobilization, the enzyme is attached by means of a covalent bond to some immobilizing group (glutaraldehyde, cyanuric chloride, etc.); in physical immobilization, the enzyme is trapped within an inert matrix or just absorbed. In using these procedures, it is important to avoid steps that would denature the enzyme. The specific immobilization scheme would depend on the particular enzyme electrode system. Such schemes permit the use of small quantities of the enzyme with large numbers of samples and offer increased pH and/or temperature stability. Hence, low-cost repetitive clinical assays are feasible. The electrode lifetime is determined primarily by the stability of the enzyme used. The lifetime can be affected by the storage conditions. Storage under refrigeration (in distilled water or buffer solution) is recommended.

The response characteristics of enzyme electrodes depend on many variables, and an understanding of the theoretical basis of their function would help to improve their performance. Enzymatic reactions involving a single substrate can be formulated in a general way as

$$E + S \underset{k_{-1}}{\overset{k_1}{\rightleftharpoons}} ES \overset{k_2}{\longrightarrow} E + P \tag{3-2}$$

In this mechanism, the substrate S combines with the enzyme E to form an intermediate complex ES, which subsequently breaks down into products P and liberates the enzyme. At a fixed enzyme concentration, the rate of the enzyme-catalyzed reaction V is given by the Michaelis-Menten equation:

$$V = V_m[S]/(K_m + [S]) \tag{3-3}$$

where K_m is the Michaelis-Menten constant and V_m is the maximum rate of the reaction. K_m corresponds to the substrate concentration for which the rate is equal to half of V_m. Figure 3-3 shows the dependence of the reaction rate upon substrate concentration, together with the parameters K_m and V_m. The initial rate increases with substrate, until a nonlimiting excess of substrate is reached, after which additional substrate causes no further increase in the rate. Hence, a leveling-off of calibration curves is expected at substrate concentrations above the K_m of the enzyme. Experimentally, the linear range may exceed the concentration correspond-

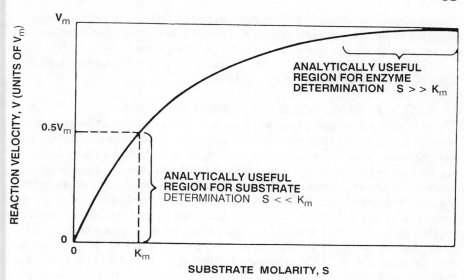

FIGURE 3-3
Dependence of the velocity of an enzyme-catalyzed reaction upon substrate concentration (at a constant level of enzyme activity).

ing to K_m, because the substrate concentration in the electrode containment region is often less than the bulk solution concentration (a common case with diffusion-limited amperometric probes). The restricted linear range of enzyme electrodes often necessitates sample dilution. Several theoretical studies of the steady-state and transient behaviors of amperometric and potentiometric enzyme electrodes have been reported.[6-8]

Amperometric-based enzyme electrodes, which consume the reaction product, display a more expanded linear range response than their potentiometric counterparts. The response time—required to reach steady-state—tends to be slow as a result of the long diffusion path between the test solution and inner detector surface. Hence, it depends on the diffusion coefficient of the substrate and the thickness of the enzyme membrane. The response time is affected also by the substrate concentration; high concentration gives a short response time due to the increased rate of the enzymatic reaction. The types of interferences common when enzyme electrodes are used include interferences associated with the measurement scheme, lack of specificity of some enzymes, and the presence of various inhibitors, e.g., heavy metal ions. Indirect measurements of inhibitors (or activators) can often be achieved by comparing the response in their presence and absence. It is possible also to determine the activity of the enzymes themselves under properly controlled conditions, including high substrate concentrations (see Figure 3-3).

In various situations, it is advantageous to immobilize more than one enzyme to the sensor surface. This is particularly the case when a single enzymatic reaction produces a nonelectroactive species, which can be broken down by a second enzyme to a detectable (electroactive) species. In addition, a considerable enhancement of the sensitivity of enzyme electrodes can be achieved by enzymatic recycling of the analyte in two-enzyme systems. Removal of interferents or side products is also feasible using the multienzyme arrangement.

The design, behavior, and utility of various enzyme electrodes of clinical significance are described in the following sections.

3-2.1.1 ENZYME FIELD EFFECT TRANSISTORS

A relatively new and promising approach for enzyme-based potentiometric probes can be obtained by immobilizing an enzyme on a field effect transistor, which results in an enzyme field effect transistor (ENFET). The enzyme is immobilized within a layer of some finite thickness which is placed on an ion-selective sensor sensitive to one of the ionic species involved in the enzymatic reaction. The advantages of ENFETs include a well-defined thickness of the gel layer, a small area that obviates the use of an external retaining membrane normally used in enzyme macroelectrodes, a small amount of the enzyme required for a probe, and the feasibility of differential measurement. ENFETs, however, suffer from the same stability problem (limited lifetime of the immobilized enzyme) that characterizes conventional enzyme electrodes. Janata and coworkers[9, 10] developed this concept, derived a mathematical treatment for the performance parameters, and verified this model experimentally. Several new ENFETs for glucose, urea, and penicillin are described in the following sections. These probes are based on a pH-sensitive field effect transistor and hence strongly depend on the buffer capacity of the sample. Future use of gas-sensing devices (e.g., ammonia-sensitive metal oxide semiconductors) should offer greater selectivity.

3-2.2 Enzyme Electrodes of Clinical Significance

3-2.2.1 GLUCOSE SENSORS

Because of the recognized diagnostic value of glucose, its routine analysis in body fluids is one of the most frequent operations in a clinical chemistry laboratory. This has fostered intense development of possible assay procedures. The glucose electrode sensor, developed by Updike and Hicks[11] (and shown in Figure 3-2), represents the first reported use of an enzyme electrode. This, and similar electrochemical probes, uses immobilized glucose oxidase and is based on the following reaction:

$$\text{D-glucose} + H_2O + O_2 \xrightarrow[\text{oxidase}]{\text{glucose}} \text{gluconic acid} + H_2O_2 \tag{3-4}$$

The consumption of oxygen or liberation of hydrogen peroxide can be monitored amperometrically with a platinum sensing probe. The hydrogen peroxide, for example, can be detected by oxidation at a platinum anode held at some positive potential:

$$H_2O_2 \rightarrow 2H^+ + O_2 + 2e^- \tag{3-5}$$

Because of the lower background current, the monitoring of hydrogen peroxide allows a more sensitive assay than the measurement of oxygen consumption. The main advantage of this approach is the high specificity of glucose oxidase to glucose. Reduced selectivity, however, may occur due to the oxidation of other oxidizable sample components. This problem can be minimized by coverage with an appropriate membrane.[12] Most commonly, a thin cellulose acetate membrane is placed in front of the platinum anode. The response time is contingent on the membrane thickness, but with a thin membrane, steady-state can be achieved within 30 s. The analytical data may be affected also by variations in the oxygen tension (as expected from Equation 3-4). This effect can be reduced by manipulation of the glucose-oxygen ratio in the enzyme layer. Such amperometric sensors can be used for the quantitation of glucose in the concentration range 10^{-7} to 10^{-3} M. (New designs yield extended linearity—up to 5×10^{-2} M—thus allowing convenient assays of undiluted blood samples.[13]) The quantitation step requires calibration with standard glucose solutions. Such amperometric schemes are employed in commercial glucose analyzers, such as those made by Yellow Springs Instrument, Leeds and Northrup, or Universal Sensors.

Other electrochemical enzymatic probes for measuring glucose have been described. Nagy et al.[14] developed a potentiometric glucose electrode based on an iodide ion-selective electrode. In this case, the hydrogen peroxide produced in the enzymatic reaction oxidized the iodide ions added to the sample solution. Changes in the level of the iodide ion were detected at the sensor surface. Similarly, a fluoride ion-selective electrode can be used to detect the fluoride ion produced in the reaction between hydrogen peroxide and an organofluorine compound.[15] A bienzyme glucose sensor, based on glucose oxidase and peroxidase, was developed to minimize interferences of oxidizable species in amperometric detection.[16] The improved selectivity was achieved by monitoring the oxidation of ferrocyanide that carries out the electron exchange between the peroxidase-active center and the electrode. Williams et al.[17] used benzoquinone instead of oxygen as the hydrogen acceptor and monitored amperometrically the liberation of hydroquinone.

Recent activity in the area of chemically modified electrodes has resulted in elegant glucose electrodes. In particular, the attachment of mediators to the surface has resulted in important advantages. For example, Cass et al.[18] immobilized a ferrocenium derivative and glucose oxidase onto a graphite electrode; the former serves as an alternate cofactor for oxygen. The reaction scheme is

$$\text{Glucose} + 2\text{FeCp}_2\text{R}^+ \xrightarrow[\text{oxidase}]{\text{glucose}} \text{gluconic acid} + 2\text{FeCp}_2\text{R} \qquad (3\text{-}6)$$

As the FeCp_2R is formed in the enzymatic reaction, it is immediately reoxidized to FeCp_2R^+ at the electrode surface. This occurs at low operating potentials where interferences due to electroactive sample constituents (e.g., ascorbic acid, uric acid, bilirubin) are minimal. In addition, the assay is insensitive to the level of oxygen in the sample. This approach is suitable for analyses of whole blood or undiluted plasma. The linear range extends from 1 to 30 nM. The concept should be widely applicable to other oxidases. Another glucose sensor, prepared by adsorption of glucose dehydrogenase and the mediator Meldola Blue on the surface of a carbon electrode, was incorporated recently with a flow injection system to allow 100 assays per hour.[19] Umana and Waller[20] described an interesting approach to trap glucose oxidase in an organic conducting film (polypyrrole); hence, the products of the enzymatic reaction are generated in the immediate proximity of the electronically conducting material. These modified-electrode-based glucose probes are indicative of future trends in substrate sensing. In particular, the ability of surface-bound mediators to shuttle electrons between enzymes and electrodes is expected to play an important role in the future development of other probes.

Instead of modifying the electrode (and using electron-transfer mediators), it is possible to modify the enzyme to enable direct electrical communication between glucose oxidase and metal electrodes.[21] The implications of this important observation on the development of glucose electrodes, particularly in-vivo ones, are obvious (see Chapter 5). Another very promising avenue is the use of organic conducting salts (e.g., tetrathiafulvalinium-tetracyanoquinodimethanide), which allow the electrooxidation of the reduced flavin redox center of glucose oxidase.[22] The electrical conductivity of these materials is associated with the delocalization of charge due to overlap of the extended π-electron systems of their molecules.

It is possible also to use soluble enzymes, instead of immobilized enzymes, for electrochemical sensing of glucose. In particular, Mottola and coworkers[23] used a closed-loop flow system with circulating glucose oxidase. Glucose samples were injected and oxygen consumption was monitored at an exposed platinum-wire electrode. Approximately 1000 samples could be processed per hour with a relative standard deviation of 2%. The same enzyme solution was used for processing more than 10,000 serum samples. The "closed-loop" concept was used also for measuring the activity of glucose oxidase.

3-2.2.2 GALACTOSE ELECTRODES

Measurements of galactose are important in the preliminary diagnosis of galactosemia and other disorders. Taylor et al.[24] described an accurate method for the determination of galactose in whole blood and plasma. An amperometric hydrogen peroxide electrode was covered by a selective membrane system containing immobilized galactose oxidase. The production of hydrogen peroxide, during the oxidation of galactose, was monitored to give a rapid (40-s) response, with minimal interference from physiologically important compounds. Cheng and Christian[25] described an amperometric method for the determination of galactose in blood serum and urine, based on the same reaction, while the enzyme is free in solution. The rate of oxygen depletion during the reaction was monitored by a membrane oxygen electrode and was shown to be directly proportional to the galactose concentration. The average recovery, with spiked serum and urine samples, was 101% over the concentration range 50–200 mg/100 ml.

3-2.2.3 UREA ELECTRODES

Various enzyme electrodes based on ion-selective or gas-selective probes have been designed to measure the physiologically important substrate urea. Both schemes rely upon the enzyme-catalyzed hydrolysis of urea:

$$(NH_2)_2 CO + 2H_2O + H^+ \rightarrow HCO_3^- + \quad 2NH_4^+ \qquad (3\text{-}7)$$
$$\Updownarrow$$
$$2NH_3 + 2H^+$$

In one case, urease is entrapped in a gel matrix that is in contact with the surface of an ammonium ion–selective electrode, which detects the ammonium ions generated.[26] Despite the good stability and dynamic response times, the sensor exhibited interferences from sodium and potassium ions. Various approaches have been suggested to minimize such interferences. In particular, Guilbault and Hrabamkova[27] treated blood and urine samples with an ion-exchange resin to reduce cationic interferences. Blaedel and Kissel[28] incorporated an anion-exchange membrane to exclude electrostatically interfering ions from the surface.

Alternatively, one could use a gas-sensing electrode for ammonia to minimize ionic interferences. For this purpose, the urease solution can be trapped between the electrode membrane and a cellophane membrane. Such an electrode was applied by Papastathopoulos and Rechnitz[29] for the determination of urea in whole blood. The response was linear in the 5×10^{-4}- to $7 \times 10^{-2}M$ concentration range. The electrode was stable for 3 weeks, after analyses of 150 samples. Chemical binding of urease to a Teflon membrane, which is an integral part of an ammonia gas membrane electrode (Figure 3-4), results in improved stability and sampling rate[30]; 20–25 serum samples can be processed in 1 h. An enzyme electrode for urea based on the ammonia air-gap electrode offers the advantages of faster response time and lower matrix effects over gas membrane sensors.[31,32] None of the common ions present in blood interferes in the assay of urea. However, the electrode assembly must be taken apart after each measurement to renew the electrolyte solution.

A pH-sensitive field effect transistor offers another promising approach for sensing urea.[33] For this purpose urease is immobilized on an aminopropyltri-ethoxysilane-glutaraldehyde membrane, held over the gate insulator. The probe had a linear range between 16 to 160 mM and a rapid (30-s) response time. Another urea electrode that offers possible miniaturization involves a coated-metal electrode (with urease coating an antimony electrode), which yields rapid pH changes in the presence of urea.[34]

3-2.2.4 AMINO ACID ELECTRODES

General amino acid electrodes using the enzyme L-amino acid oxidase in conjunction with potentiometric or amperometric sensing probes have been reported.[35–37] The enzyme catalyzes the reaction:

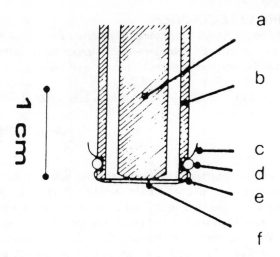

FIGURE 3-4
Ammonia gas membrane electrode: (a) glass electrode; (b) electrode outer jacket; (c) Teflon membrane; (d) O-ring; (e) epoxy seal; (f) enzyme chemically bound and attached to membrane. (Reproduced with permission.[30])

$$RCHNH_3^+COO^- + H_2O + O_2 \xrightarrow{\text{L-amino acid oxidase}}$$

$$RCOCOO^- + NH_4^+ + H_2O_2 \qquad (3\text{-}8)$$

The ammonium ion produced can be detected at a monovalent cation electrode; a dialysis membrane is used to hold the enzyme on the surface. The degree of response varies with the amino acid tested, because of variation in the activity of the enzyme with the acid. Rather than the cation-selective electrode, it is possible to use amperometric detection of the generated hydrogen peroxide at carbon or platinum anodes.[16, 36] In particular, immobilization of L-amino acid oxidase to a graphite electrode (via a cyanuric chloride linkage) resulted in a rapid (25-s) response and an expanded linear range.[37] The useful lifetime of this electrode was 30 days.

More selective amino acid electrodes have been reported using specific deaminase and decarboxylase enzymes. For example, Guilbault and Shu[38] described an electrode for L-tyrosine based on the enzyme tyrosine decarboxylase and a carbon dioxide sensor (that senses the liberation of this gas). A similar approach can be applied for measuring L-lysine.[39] Direct immobilization of lysine decarboxylase onto the surface of a carbon dioxide sensor yielded a stable and specific response. Fung et al.[40] described a potentiometric probe for the determination of methionine. The enzyme methionine lyase was immobilized on an ammonia probe. The enzyme was stable for 3 months and yielded a linear response over the range 10^{-2} to 10^{-5} M. An L-asparagine-specific enzyme electrode, produced by immobilizing the enzyme asparaginase to an ammonia sensor, yielded a linear response over the 8×10^{-5}- to $8 \times 10^{-3} M$ range.[41] A selective phenylalanine electrode, based on the enzyme L-phenylalanine ammonia lyase and an ammonia air-gap electrode, was described by Hsiung et al.[42] A highly selective histidine electrode was prepared by immobilizing histidine decarboxylase at the surface of a potentiometric carbon dioxide sensor.[43] The electrode had an average slope of 50 mV/decade over a range from 3×10^{-4} to 1×10^{-2} M, and allowed reliable determination of histidine in urine. A tyrosine-selective electrode was applied for a reliable and simple determination of total protein in serum.[44] The protein in serum was first hydrolyzed with pepsin, and the tyrosine cleaved was subsequently determined. A very good agreement with the biuret method was obtained.

3-2.2.5 CREATININE ELECTRODES

The determination of creatinine in body fluids is of significant value for diagnosis of renal, muscular, and thyroid function. An ammonia gas-sensing electrode has been used with immobilized creatininase to measure creatinine levels.[45] The enzymatic formation of ammonia from creatinine was monitored at the optimum pH of the enzyme (8.5). Ammonia-sensitive semiconductor probes (based on iridium-metal oxide) can also be used to monitor the same enzymatic reaction and to rapidly and reliably determine creatinine in body fluids.[46] A novel multienzyme membrane electrode system for the determination of creatinine and creatine in serum was described by Tsuchida and Yoda.[47] The three enzymes (creatinine amidohydrolase, creatine amidinohydrolase, and sarcosine oxidase) were coimmobilized onto the porous side of a cellulose acetate membrane that has selective permeability to hydrogen peroxide. Amperometric detection of the latter resulted in a linear response up to 100 mg/l of creatinine and creatine in serum, with a detection limit of 1 mg/l. Only 25 μl of serum was required, the response time was 20 s, and the system was sufficiently stable to perform more than 500 assays. Danielsson et al.[48] described a palladium-coated semiconductor probe for enzymatic analysis of creatinine. Creatinine deiminase was immobilized in the enzyme thermistor unit, and the ammonia generated was monitored.

3-2.2.6 PENICILLIN ELECTRODES

The enzyme penicillinase has been employed in the construction of an electrode for penicillin.[49] The enzyme was immobilized in a polyacrylamide gel on a glass pH electrode. The decrease in pH due to the production of penicilloic acid was sensed by the glass electrode, and the potential change was proportional to the logarithm of the penicillin concentration. Best results were obtained around the optimum pH (6.4) for the enzyme, yielding a response time less than 30 s. An improved penicillin-selective enzyme electrode, based on immobilizing penicillinase by adsorption onto a fritted glass electrode, was shown to minimize interferences due to monovalent cations and to offer improved stability.[50] Caras and Janata[51] described a penicillin-sensitive field effect transistor. The probe was constructed by depositing a co-cross-linked penicillinase-albumin layer over a pH-sensitive field effect transistor. It had a lifetime of 2 months, rapid response time, and sensitivity and linear range comparable to analogous enzyme macroelectrodes. Owing to its small size, only a minute amount of enzyme was required. Differential operation provided appropriate temperature compensation.

3-2.2.7 URIC ACID ELECTRODES

The enzyme uricase, in conjunction with a carbon dioxide gas sensor, was used by Kawashima and Rechnitz[52] to construct an electrode for uric acid. The enzyme oxidizes uric acid according to the reaction

$$\text{Uric acid} + O_2 \xrightarrow{\text{uricase}} \text{allantoin} + H_2O_2 + CO_2 \qquad (3\text{-}9)$$

The sensor gave a linear response from 1×10^{-4} to $2.5 \times 10^{-3}\ M$, with a slope of 57 mV/decade. It is possible also to monitor the disappearance of dissolved oxygen during the enzymatic degradation of uric acid by using a platinum electrode covered with the immobilized enzyme.[53] By this method, rapid assays of uric acid in human serum and urine were obtained.

3-2.2.8 OXALATE ELECTRODES

Guilbault and coworkers[54] described an enzyme electrode for the determination of oxalate in urine. The electrode was based on the immobilization of the enzyme oxalate oxidase on pig intestine mounted on the tip of an oxygen electrode. The enzyme catalyzes the reaction

$$(COOH)_2 + O_2 \xrightarrow{\substack{\text{oxalate} \\ \text{oxidase}}} 2CO_2 + H_2O_2 \qquad (3\text{-}10)$$

Changes in the oxygen partial pressure were monitored amperometrically. The method was applied to small volumes of untreated urine and yielded good agreement with an established spectrophotometric method.

3-2.2.9 CHOLESTEROL ELECTRODES

The determination of cholesterol in body fluids has long been of interest because of its medical importance. Hahn and Olson[55] described an amperometric enzymatic method for the determination of total cholesterol in serum. Three enzymes were used (cholesterol esterase, cholesterol oxidase, and peroxidase) to liberate free cholesterol, oxidize it, and convert the resulting hydrogen peroxide into equivalents of ferricyanide which is reduced at a tubular carbon electrode. Good sensitivity, accuracy, and reproducibility were reported. Another amperometric determination of total cholesterol in serum, using immobilized cholesterol ester hydrolase and cholesterol oxidase in conjunction with a platinum anode to detect the hydrogen peroxide liberated enzymatically, was described by Guilbault and coworkers.[56] The calibration curve for total serum cholesterol was linear from 0 to 5.0 g/l. The enzyme layer was placed on a rotating porous cell that was positioned close to the electrode. Amperometric monitoring of the rate of oxygen depletion in the cholesterol oxidase–catalyzed oxidation of cholesterol can also be used for measuring total cholesterol in plasma.[57] A similar procedure was improved by Clark et al.[58] to allow 1-min assays in matrices such as plasma, serum, or tissues. Ascorbic acid, uric acid, and bilirubin did not interfere. An excellent correlation with the Abell-Kendall colorimetric method was reported.

3-2.2.10 ETHANOL ELECTRODES

The determination of ethanol in blood is commonly performed in clinical and forensic laboratories. Ethanol may be made to react with the cofactor nicotinamide adenine dinucleotide (NAD^+) in the presence of the enzyme alcohol dehydrogenase (ADH):

$$C_2H_5OH + NAD^+ \xrightarrow{ADH} C_2H_5O + NADH \qquad (3\text{-}11)$$

This reaction serves as a basis of various amperometric sensors for ethanol in which ADH is immobilized to various platinum or carbon anodes.[59–61] The enzymatically produced NADH is monitored in the following reaction:

$$NADH \rightarrow NAD^+ + 2e^- + H^+ \qquad (3\text{-}12)$$

The oxidation current is taken as an estimate of the ethanol content of the sample. The oxidation reaction serves also to regenerate the oxidized cofactor, which can be used for further determination (Figure 3-5). An acetylated dialysis membrane can be used to constrain both the cofactor and the enzyme.[60]

Alternatively, the NADH may be aerobically oxidized by molecular oxygen in the presence of horseradish peroxidase, with the rate of oxygen depletion (as monitored with an oxygen-sensing electrode) being directly proportional to the

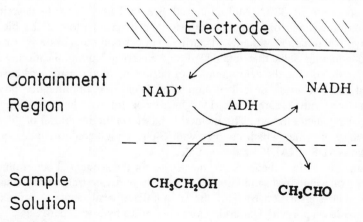

FIGURE 3-5
Reagentless ethanol electrode.

ethanol concentration.[62] The latter scheme can use untreated whole blood samples with volumes as small as 5 μl. Response time was less than 60 s, and correlations with standard methods were very good. Smith and·Olson[63] used a differential amperometric procedure for the determination of ethanol in blood and urine. The NADH produced in reaction 3-11 was coupled to the redox mediator and electroactive species 2,6-dichlorophenolindophenol. The amperometric detection of the mediator at a tubular carbon electrode eliminated possible interferences; measurements were made on 10–20-μl samples.

3-2.2.11 LACTATE ELECTRODES

The electrocatalytic oxidation of reduced nicotinamide coenzymes is useful for measuring other substrates, the reactions of which are catalyzed by dehydrogenases. Blaedel and Jenkins[64] developed a "reagentless" electrode for lactate. In their scheme, lactate dehydrogenase and NAD^+ were trapped in a thin layer between the glassy carbon electrode and a cellulose membrane; the latter permits transport of lactate (in) and pyruvate (out) while retaining the enzyme and coenzyme (in a design similar to that shown in Figure 3-5). The enzymatic reaction of lactate dehydrogenase can also proceed in the presence of the ferricyanide-ferrocyanide redox couple.[65] Lactate is being oxidized by ferricyanide, and the liberation of ferrocyanide is monitored at a platinum anode. Other lactate dehydrogenase electrodes were described, including an ultramicroelectrode that allowed the determination of pyruvate in a rat cerebrospinal sample of 50 μl,[66] and a modified glassy carbon probe used to determine the activity of the enzyme in biological fluids and tissues.[67]

3-2.2.12 MISCELLANEOUS ENZYME ELECTRODES

Deng and Enke[68] developed an adenosine-selective electrode by coupling the enzyme adenosine deaminase with an ammonia gas-sensing membrane electrode. The electrode allowed detection of micromolar levels of adenosine using a pH of 9 and a temperature of 37°C. McKenna and Brajter-Toth[69] recently described an amperometric xanthine oxidase electrode for the determination of the biological purines hypoxanthine, xanthine, and purine. The working electrode was made of organic conducting salts that enabled direct electrical communication with the enzyme. Application to the determination of purine in blood plasma was described. Seegopaul and Rechnitz[70] described the use of an enzyme amplification procedure in conjunction with a carbon dioxide gas sensor for the determination of the anticancer drug methotrexate. The method is based on the inhibition of dihydrofolate reductase enzyme, which couples with 6-phosphogluconic dehydrogenase to recycle the $NADP^+/NADPH$ redox couple.

Kihara et al.[71] described a bienzyme sensor for the sequential determination of glutamate-oxalacetate transaminase and glutamate-pyruvate transaminase activities in serum. The sensor consisted of the immobilized oxalacetate decarboxylase–pyruvate oxidase–polyvinyl chloride membrane and a hydrogen peroxide electrode. An enzyme electrode for salicylate, based on trapping salicylate hydroxylase at the sensing tip of a carbon dioxide probe, was described by Fonong and Rechnitz.[72]

The determination of salicylate in pooled serum was demonstrated, with a good correlation to a standard spectroscopic method. Baum[73] described an enzyme electrode to measure acetylcholine. A Nernstian response was obtained in the $10^{-1}-10^{-5}M$ range, with good selectivity over choline, or sodium, potassium, and ammonium ions.

3-2.2.13 ARTIFICIAL ENZYME–BASED ELECTRODES

Ho and Rechnitz[74] reported recently the properties of a new type of electrochemical biosensor based on an artificial enzyme. Artificial enzymes are synthetic polymer chains that have functional groups that mimic the biocatalytic activity of natural enzymes. The main advantage in coupling artificial enzymes with biosensors is long-term stability. In addition, artificial enzymes have no cofactor requirement and exhibit a broad pH profile.

3-2.3   Conclusion

The survey of the enzyme electrodes given here (and summarized in Table 3-1) shows the wide range of substances for which electrodes have been constructed so far. Their full potential, however, has not yet been realized. As the number of enzymes that have been isolated and purified grows and our knowledge of these catalysts increases, enzyme electrodes will play an even greater role in clinical chemistry. In particular, a proper solution to the stability problem will undoubtedly lead to wide-scale commercialization.

3-3   SENSORS BASED ON OTHER BIOCATALYTIC LAYERS

3-3.1   Introduction

The use of tissue slices or bacterial cells of animal or plant origin as immobilized biocatalysts has extended the scope of electrochemical sensors beyond that of conventional enzyme electrodes. These sensors function in a manner similar to that of conventional enzyme electrodes, i.e., detection of an electroactive species that is produced or consumed by the enzyme present in the tissue or cell. However, in several cases such biocatalysts have been shown to possess important advantages, including improved stability, lower cost, and higher activity, compared with the use of isolated enzymes. In addition, required cofactors may already be present in the cell and do not need to be separately immobilized.

Such electrodes may be particularly attractive when isolated enzymes are not readily available or when multistep reaction paths are required. The stability advantage is illustrated in Figure 3-6, which compares the useful lifetimes of various biocatalytic electrodes used for measuring glutamine. (An appropriate preservative is often added to the working buffer when animal tissue electrodes are concerned.) Extended lifetimes of microbial electrodes can be achieved via an

Table 3-1
Some Common Enzyme Electrodes

Measured Species	Enzyme	Detected Species	Type of Sensing Probe	Reference
Adenosine	Adenosine deaminase	NH_3	Potentiometric gas sensing	68
Cholesterol	Cholesterol oxidase, cholesterol esterase	$K_4Fe(CN)_6$	Amperometric	55
Creatinine	Creatinase	NH_3	Potentiometric gas sensing	45
Ethanol	Alcohol dehydrogenase	NADH	Amperometric	59
Ethanol	Alcohol dehydrogenase, horseradish peroxide	O_2	Amperometric	62
Galactose	Galactose oxidase	O_2	Amperometric	25
Glucose	Glucose oxidase	O_2, H_2O_2	Amperometric	11
Glucose	Glucose oxidase	Ferrocenium	Amperometric	18
Glucose	Glucose oxidase	F^-	Potentiometric	15
Lactate	Lactate dehydrogenase	NADH	Amperometric	64
Lysine	Lysine decarboxylase	CO_2	Potentiometric gas sensing	39
Methionine	Methionine lyase	NH_3	Potentiometric gas sensing	40
Oxalate	Oxalate oxidase	O_2	Amperometric	54
Penicillin	Penicillinase	H^+	Potentiometric	49
Penicillin	Penicillinase	H^+	Field effect transistor	51
Purine	Xanthine oxidase	Xanthine oxidase	Amperometric	69
Salicylate	Salicylate hydroxylase	CO_2	Potentiometric gas sensing	70
Tyrosine	Tyrosine decarboxylase	CO_2	Potentiometric gas sensing	36
Urea	Urease	NH_4^+	Potentiometric gas sensing	26
Urea	Urease	NH_3	Potentiometric gas sensing	29
Uric acid	Uricase	CO_2	Potentiometric gas sensing	52

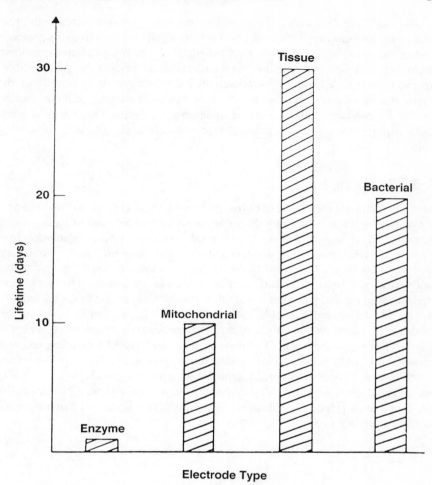

FIGURE 3-6
Comparison of minimum useful lifetimes of various glutamine
biosensors. (Reproduced with permission.[75])

in-situ regeneration of the biocatalytic activity, accomplished by placing the
electrode in a nutrient growth medium to grow a fresh supply of cells. Tissue
electrodes, however, are easier to construct compared with bacterial ones, because no cultivation is required. The animal tissue is usually physically retained
at the surface of the sensing probe using support membranes (often in a sandwich configuration between inner and outer membranes). The optimum thickness
of the tissue slice is often a compromise between mechanical stability and response
time.

Of particular interest is the recent use of plant tissues, which opens many
possibilities. Such tissues were thought of as having less biocatalytic activity than
animal tissues. Many plant tissues, however, have been shown recently to exhibit
high levels of activity and are highly suited for biosensing purposes.

The major problem with this type of biocatalytic sensor (compared with those based on isolated enzymes) has been lower selectivity. This arises from the presence of other enzymes, which result in multiple metabolic pathways. Improvement in selectivity, via a judicious "tuning" of the biochemical process, is often possible. This can be accomplished by appropriately treating the tissue or by controlling the bacteria during its growth phase. In addition, unwanted enzyme activities can be inhibited by choosing suitable enzyme inhibitors.[76] Nevertheless, potential users should explore the selectivity properties of each electrode to avoid misleading results.

3-3.2 Examples

One of the first (and more selective) tissue electrodes was constructed by holding a slice of porcine kidney (0.5 mm thick) at the surface of an ammonia gas-sensing electrode.[77] This sensor displayed very good selectivity toward glutamine over potential interferences, with good sensitivity and response slope. Smit and Rechnitz[78] described a leaf-based biosensor for L-cysteine. A leaf from a cucumber plant was immobilized at the surface of an ammonia gas sensor. The outer waxy cuticle layer of the leaf was removed to allow the substrate access to the biocatalytic activity. Macholan and Chmelikova[79] described an amperometric biosensor for ascorbic acid, based on coupling a slice of the mesocarp of squash or cucumber to a Clark-type oxygen electrode. One tissue slice served up to 80 measurements, with a response time of 80 s and a relative standard deviation of 3%. Tissue portions from various flowers have been coupled recently to an ammonia gas electrode for use as urea and glutamine sensors.[80, 81] Different patterns and response characteristics were found among species of flowers. The construction of such a magnolia petal biosensor is shown in Figure 3-7.

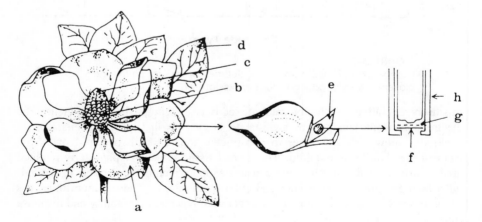

FIGURE 3-7
Schematic diagram of a magnolia flower and construction of a biosensor. (a) Petal; (b) stamen; (c) carpel; (d) leaf; (e) petal tissue slice; (f) tissue disk; (g) gas-permeable membrane; (h) electrode body. (Reproduced with permission.[81])

Updike and Treichel[82] described a tissue electrode responsive to an antidiuretic hormone. A toad bladder tissue membrane was held at the surface of a sodium ion–sensing glass electrode. The hormone enhanced the transport of sodium ions across the tissue, and the resulting potentials were proportional to the hormone concentration. Interference from other hormones affecting the sodium transport was reported. A dopamine-specific tissue probe was constructed by placing banana slices on the gas-permeable membrane of an oxygen electrode.[83] The banana slices contained polyphenol oxidase, which catalyzed the reaction of dopamine and oxygen to form dopamine quinone. A useful lifetime in excess of 1 week was reported. A mixed banana–carbon paste electrode (with an intimate contact between the biocatalytic and sensing sites) offered the advantage of very short response times.[84] Arnold and Rechnitz[85] described a rabbit muscle–based potentiometric probe for adenosine 5'-monophosphate (AMP). The slice of rabbit muscle was immobilized at the surface of an ammonia gas-sensing electrode, which monitored the ammonia produced in the biocatalytic process. The probe had a slope of 57 mV per AMP concentration decade over the linear range, which extended from 1.4×10^{-4} to $1.0 \times 10^{-2} M$. A useful lifetime of 4 weeks was reported. The biocatalytic activity was much greater than that of the comparable enzyme electrode. These same researchers developed a rabbit liver–based potentiometric probe for guanine.[86] The thin tissue slice was held at the surface of an ammonia gas-sensing electrode. The development of this guanine biosensor included a step-by-step optimization strategy to affect the substrate diffusion and biocatalytic activity. The response to guanine at various electrode ages (Figure 3-8) indicates the inherent stability of this sensor.

Many bacterial electrodes have utilized bacteria that have selective deaminase activity in conjunction with an ammonia-selective electrode. For example, a bacterial electrode for measuring arginine was described by Rechnitz et al.[87] A layer of Streptococcus faecalis was fixed to the surface of an ammonia gas probe. This layer contained the enzyme arginine deaminase that catalyzed the transformation of L-arginine to citrulline and ammonia, which is then sensed by the internal probe. Response to arginine was observed between 5×10^{-5} and $1 \times 10^{-3} M$, with approximately 15 min required to obtain the steady-state potential. The fact that the enzyme arginine deaminase is not available commercially represents an obvious advantage of this probe. Karube et al.[88] described an amperometric microbial sensor for ammonia, consisting of immobilized nitrifying bacteria and an oxygen electrode. A detection limit of 0.1 mg/l and response time of 4 min were reported. The same group[89] described a microbe-based electrochemical system for rapid screening of chemical mutagens. The system was composed of an oxygen electrode and a filter paper retaining Salmonella typhihimurium revertant. The current decrease of this sensor in a glucose-containing buffer, due to respiration of the bacterial cells, was correlated with mutagen concentrations. The 10-h period required to complete the screening is substantially shorter than the 2 days of incubation needed for the conventional mutagen test.

Di Paolantonio and Rechnitz[90] described a bacterial tyrosine-selective potentiometric electrode in which the desired biocatalytic activity was induced biochemically during growth of the bacterial cell. The normally ineffective

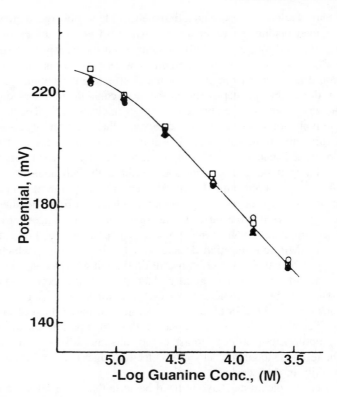

FIGURE 3-8
Response of the guanine electrode at various electrode ages: day 3
(○), day 6 (□), day 10 (Δ), and day 14 (●). (Reproduced with
permission.[86])

biocatalyst *Aeromonas phenologenes ATCC 29063* was coupled with an ammonia
gas-sensing electrode. The biochemical induction resulted in a sharp increase of
biocatalytic activity without loss of selectivity. The probe had an average slope of
54 mV per concentration decade over a range from 8.3×10^{-5} to 1.0×10^{-3} M
tyrosine and a detection limit of 3.3×10^{-5} M tyrosine. The same authors
reported[91] a bacteria-based potentiometric probe for pyruvate. *Streptococcus
faecium*, containing pyruvate dehydrogenase, served to convert pyruvate to carbon
dioxide, which was detected by a gas-sensing electrode. The sensor had a slope of
41 mV per concentration decade over a range from 2.2×10^{-4} to 3.2×10^{-2} M
pyruvate.

Karube and coworkers[92] described a microbial potentiometric probe for
measuring the antibiotic cephalosporin. The enzyme cephalosporinase, contained in
the *Citrobacter freundii* used, catalyzed the reaction of cephalosporin to generate
hydrogen ions, which were sensed at the glass pH electrode. Other antibiotics can
be measured based on their effect on the metabolic processes of sensitive
microorganisms. For example, Simpson and Kobos[93] described an assay for

tetracycline based on the inhibition of carbon dioxide production of an *E. Coli* suspension.

3-4 ELECTROCHEMICAL IMMUNOSENSORS

Immunological reactions provide another promising route to the operation of electrochemical biosensors. The development of electrochemical immunosensors is a consequence of the need for greater specificity. In particular, enzyme immunosensors based on the competitive or sandwich modes of operation are of great current interest because of their applicability to a wide range of analytes of clinical significance.

The principle of operation is illustrated in Figure 3-9. In order to determine the antigen, the corresponding antibody is immobilized on a membrane matrix, which is held on an amperometric or potentiometric sensing probe (e.g., oxygen or iodide electrodes) used to measure the rate of the enzymatic reaction. In the competitive binding assay, the sample antigen competes with enzyme-labeled antigen for the antibody-binding sites on the membrane. The membrane is then washed, and the probe is placed in a solution containing the substrate for the enzyme. For example, Aizawa et al.[95] developed an enzyme immunosensor for the electrochemical determination of the tumor antigen α-fetoprotein. The sensor was prepared by attaching the antibody-containing membrane to a Clark-type oxygen probe (Figure 3-10). The enzyme catalase was used as a labeling agent. After competitive binding, the sensor was examined for the catalase activity via its effect on the decomposition of hydrogen peroxide to release oxygen. Similar enzyme immunosensors—based

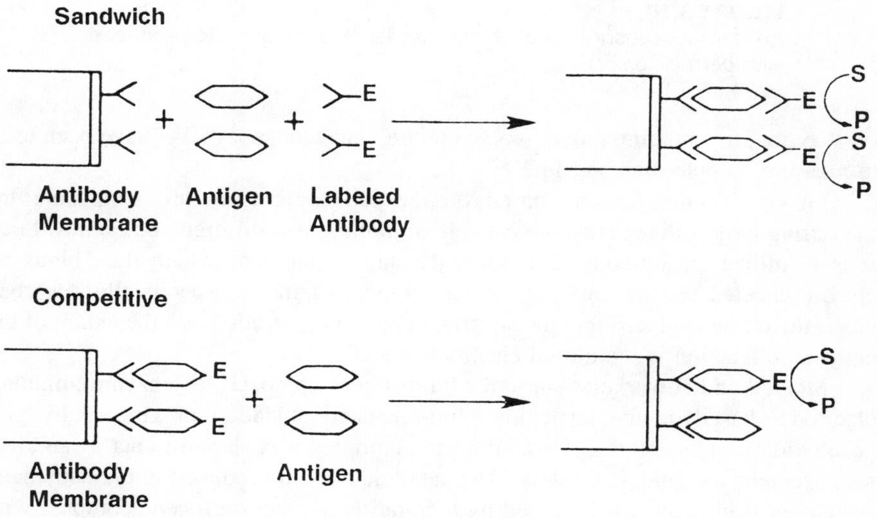

FIGURE 3-9
Enzyme immunosensors. (Reproduced with permission.[94])

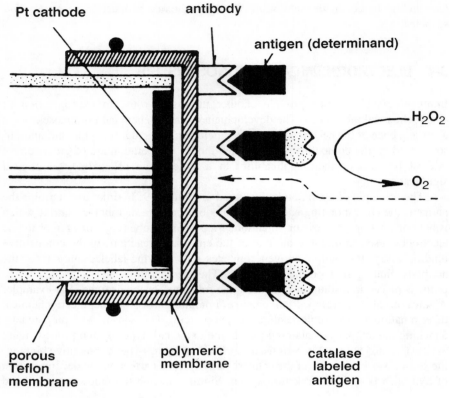

FIGURE 3-10
An immunosensor based on catalase-labeled antigen. (Reproduced with permission.[95])

on a competitive binding mode, oxygen probe, and catalase label—have been used to measure theophylline and IgG.[94]

Enzyme immunosensors based on the sandwich assay are applicable for measuring large antigens that are capable of binding two different antibodies. Such sensors utilize an antibody that binds the analyte-antigen, which then binds an enzyme-labeled second antibody. After removal of the nonspecifically adsorbed label, the probe is placed into the substrate-containing solution and the extent of the enzymatic reaction is monitored electrochemically.

McNeil and coworkers[96] described a different approach for enzyme immuno-electrodes based on the ferricinium ion-mediated oxidation of glucose by glucose oxidase. The antigen–ferricinium ion complex was shown to act as an electron acceptor for glucose oxidase. The catalytic current produced in the enzymatic oxidation of glucose was inhibited by binding the antigen-ferrocene complex with antibody. The inhibition could be reversed by adding the free antigen. Measurement of the drug lidocaine in plasma was used to illustrate this scheme.

Keating and Rechnitz[97] developed another class of immunoelectrode based on the coupling of antibody-antigen reactions to potentiometric membrane electrodes (Figure 3-11). They term this electrode system a *potentiometric ionophore-modulated immunoassay* (PIMIA), because the electrode response involves the interaction between an antibody in solution with an antigen-ionophore complex in

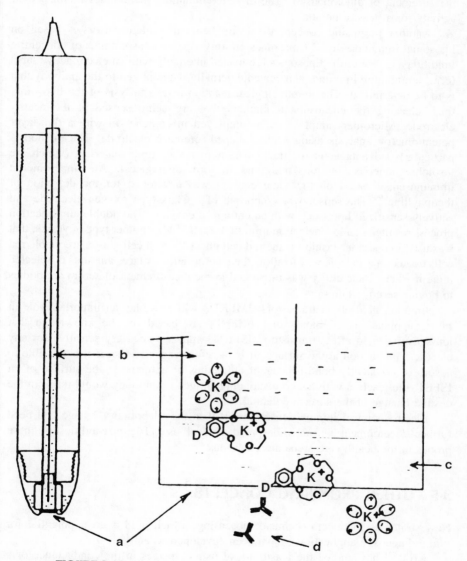

FIGURE 3-11
A digoxin antibody–sensing electrode: (a) PVC membrane containing digoxin-carrier conjugate, (b) inner filling solution, (c) plasticizer, (d) digoxin antibodies. (Reproduced with permission.[97])

a plasticized polymer membrane. A difference in membrane potential is measured upon antibody binding to the membrane-bound antigen-ionophore complex. This immunosensor was developed for the selective measurement of the antibodies to the cardiac drug digoxin in the milligrams-to-liter range. (The digoxin-dibenzo-18-crown-6 was used as the complex.) This competitive approach also allowed trace measurement of digoxin itself. The digoxin-containing membranes retained their activity for at least a month.

Another promising concept to electrochemical immunosensors is based on liposomal immunoassay.[98] Liposomes are tiny, spherical assemblies of concentric lipid bilayers. Here, the liposomes are loaded internally with an electroactive label (e.g., ammonium ion) and then coated externally with antigen to the antibody that is to be determined. The immunological reaction results in lysis of the liposome, thus releasing the encapsulated markers that are being sensed at the nearby electrode. Microliter samples can be employed in conjunction with a thin-layer potentiometric analysis using a plate-shaped reference electrode. The method is particularly advantageous for trace measurements of lipid materials (which are insoluble aqueous solutions and tend to form aggregates). An amperometric immunosensor based on the same concept was developed for the detection of theophylline.[99] The sensor was composed of a Clark-type oxygen electrode and actively sensitized liposome, with an entrapped enzyme. The double amplification process resulted in a low detection limit of $4 \times 10^{-9} M$. Another type of vesicle that was used in a similar fashion is the red cell ghost.[100] Red cell ghosts are viable red cells treated in a specific way to allow the incorporation of large amounts of selected ionic marker. Their utility was illustrated in the measurement of serum antibodies to bovine serum albumin.

Immuno field effect transistors (IMFETs) will also play an important role in future applications of biosensors. IMFETs are based on the knowledge that ion-sensitive field effect transistors (ISFETs) appear to be very sensitive for any electrical interaction at the surface. It is therefore expected that if it is possible to adsorb or covalently bond a layer of antibodies or antigens to the surface of an ISFET, the reaction with the corresponding antigen or antibody would result in the desired change of the surface potential.

Electrochemical immunosensors are still in the "laboratory" stage and need further development before widespread clinical use. The possibilities of these probes in the health-care arena are enormous.

3-5 OTHER PROMISING CONCEPTS

New strategies for electrochemical biosensing, which hold a great potential for future diagnostic applications, have been developed recently.

Krull[101] has studied the feasibility of using changes in the conductance of a bilayer lipid membrane (BLM), caused by selective interaction of analyte with a membrane-embedded receptor, as a basis for a selective biosensing system. The BLM is well known as a synthetic model for natural cell membranes. Hence, sensing low levels of biomolecules may be possible via a careful "engineering" of the BLM to contain selective binding elements (e.g., antibodies, ionophores, enzymes). This work demonstrates that thin lipid bilayer membranes are not as fragile as one might expect and may provide very useful information. The prospects of lipid membrane biosensors were reviewed by Krull and Thompson.[102]

It is possible also to exploit changes in the membrane potential due to selective displacement reactions for effective sensing. For example, Yao and Rechnitz[103] developed a potentiometric biosensor for riboflavin based on the use of aporiboflavin-binding protein (Figure 3-12). This device exhibits good response characteristics for the selective determination of riboflavin in the 0.1- to $2\mu M$ range, with rapid and reproducible response.

Ikariyama et al.[104] recently developed a unique electrochemical biosensor based on bioaffinity differences which used a metastable molecular complex receptor and enzyme amplifier (Figure 3-13). Biotin was measured as it displaced

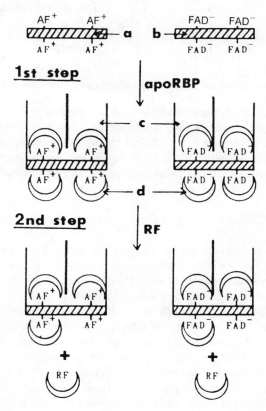

FIGURE 3-12
Potentiometric sensor for riboflavin: (a) acriflavine (AF$^+$) bound membrane; (b) flavin adenine dinucleotide (FAD$^-$) bound membrane; (c) internal solution; (d) aporiboflavin-binding protein (apoRBP); riboflavin (RF). The first step is the preparation of apoRBP-complexed membranes, and the second step is the riboflavin sensing. (Reproduced with permission.[103])

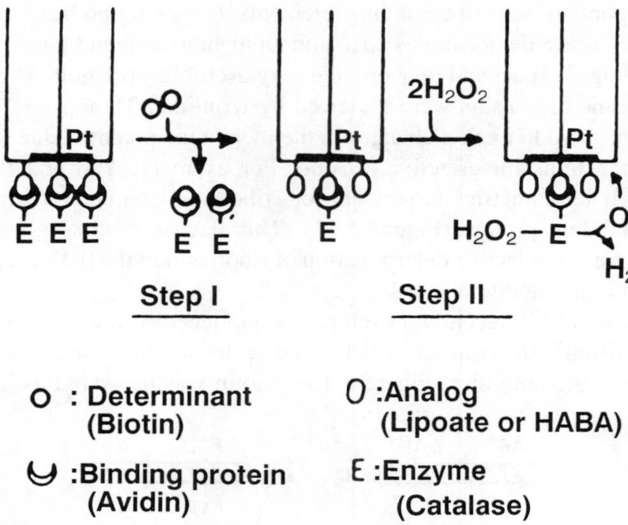

Step I   **Step II**

○ : **Determinant**   ⦰ :**Analog**
     (Biotin)   **(Lipoate or HABA)**

∪ :**Binding protein**   Ε :**Enzyme**
     (Avidin)   **(Catalase)**

FIGURE 3-13
Bioaffinity sensor for biotin. (Reproduced with permission.[104])

4-(hydroxyphenylazo)-benzoic acid from its complex with avidin. Catalase was used as an enzyme amplifier, and a Clark-type oxygen electrode was used as a sensing probe. The sensitization achieved by the enzyme amplification technique allowed measurements in the $10^{-7}-10^{-9}$-mg/l range. This concept is applicable to the determination of other small molecules, such as drugs or hormones.

3-6 CONCLUSION

We are continually witnessing the exciting introduction of new electrochemical sensing devices, as natural materials—displaying molecular-level recognition capabilities—are being coupled to amperometric or potentiometric probes. In view of the wide range of biologically active materials that might be used, many other biocomponent-electrode combinations could clearly be envisaged. Very new sensing principles, such as the use of bilayer lipid membranes, bioaffinity probes, neuroreceptors, and immuno field effect transistors, have already demonstrated a great potential. Some of these developments are so new that it is often difficult to predict future directions. Electrochemical biosensors will undoubtedly become more sophisticated in terms of physiological and biochemical schemes, and thus will be applicable to even more challenging clinical problems. New biotechnological processes, electrode modification schemes, and solid-state fabrication technology will all contribute further to sensor development. The future is almost certain to see attractive biosensing devices and faster commercial development. Clearly, electrochemical biosensors will play an increasingly significant role in all aspects of health care. The continuous signal provided by many electrochemical biosensors may be exploited for continuous in-vivo monitoring (see Chapter 5). More information on electrochemical biosensors can be found in several survey papers[93, 105–108] and a recent book.[109]

REFERENCES

1. Bower, L. D., Carr, P. W., *Anal. Chem.*, *48*, 545A (1976).
2. Zaborsky, O. R., *Immobilized Enzymes*, CRC Press, Cleveland, Ohio, 1973.
3. Weetall, H. H., *Anal. Chem.*, *46*, 602A (1974).
4. Gray, D. N., Keyes, M. H., Watson, B., *Anal. Chem.*, *49*, 1067A (1977).
5. Razumas, V. J., Jasaitis, J. J., Kulys, J. J., *Bioelectrochem. Bioenerg.*, *12*, 297 (1984).
6. Mell, L. D., Maloy, J. T., *Anal. Chem.*, *47*, 299 (1975).
7. Tran-Minh, C., Broun, G., *Anal. Chem.*, *47*, 1359 (1975).
8. Carr, P. W., *Anal. Chem.*, *49*, 799 (1977).
9. Caras, S. D., Janata, J., Saupe, D., Schmitt, K., *Anal. Chem.*, *57*, 1917 (1985).
10. Caras, S. D., Petelenz, D., Janata, J., *Anal. Chem.*, *57*, 1920 (1985).
11. Updike, S. J., Hicks, G. P., *Nature*, 214, 986 (1967).
12. Palleschi, C., Rahni, M. A. N., Lubrano, G. J., Ngwainbi, J. N., Guilbault, G. G., *Anal. Biochem.*, *159*, 114 (1986).
13. Mullen, W. H., Keedy, F. H., Churchouse, S. J., Vadgama, P. M., *Anal. Chim. Acta*, *183*, 59 (1986).
14. Nagy, G., von Storp, L. H., Guilbault, G. G., *Anal. Chim. Acta*, *66*, 443, (1973).
15. Siddiqi, I. W., *Clin. Chem.*, *28*, 1962 (1982).
16. Kulys, J. J., Pesliakiene, M. V., Samalius, A. S., *Bioelectrochem. Bioenerg.*, *8*, 81, (1981).
17. Williams, D. L., Doirg, A., Korosi, A., *Anal. Chem.*, *442*, 118 (1970).
18. Cass, A. E. G., Davis, G., Francis, G. D., Hill, A. O., Aston, W. J., Higgins, I. G., Plotkin, E. V., Scott, L. D. L., Turner, A. P. F., *Anal. Chem.*, *56*, 667 (1984).
19. Marko-Varga, G., Appelquist, R., Gorton, L., *Anal. Chim. Acta*, *179*, 371 (1986).
20. Umana, M., Waller, J., *Anal. Chem.*, *58*, 2979 (1986).
21. Degani, Y., Heller, A., *J. Phys. Chem.*, *91*, 1285 (1987).
22. Albery, W., Bartlett, P. N., Craston, D. H., *J. Electroanal. Chem.*, *194*, 223 (1985).
23. Wolff, C. M., Mottola, H. A., *Anal. Chem.*, *50*, 94 (1978).
24. Taylor, P. J., Kmetec, E., Johnson, J. M., *Anal. Chem.*, *49*, 789 (1977).
25. Cheng, F. S., Christian, G. D., *Anal. Chim. Acta*, *104*, 47 (1979).
26. Guilbault, G. G., Montalvo, J. G., *J. Am. Chem. Soc.*, *92*, 2533 (1970).
27. Guilbault, G. G., Hrabankova, E., *Anal. Chim. Acta*, *52*, 287 (1970).
28. Blaedel, W. J., Kissel, T. R., *Anal. Chem.*, *47*, 1602 (1975).
29. Papastathopoulos, D. S., Rechnitz, G. A., *Anal. Chim. Acta*, *79*, 17 (1975).
30. Mascini, M., Guilbault, G. G., *Anal. Chem.*, *49*, 795 (1977).
31. Ruzicka, J., Hansen, E. H., *Anal. Chim. Acta*, *69*, 129 (1974).
32. Guilbault, G. G., Tarp, M., *Anal. Chim. Acta*, *73*, 355 (1974).
33. Karbue, I., Tamiya, E., Dicks, J. M., Gotoh, M., *Anal. Chim. Acta*, *185*, 195 (1986).
34. Alexander, P. W., Joseph, S. P., *Anal. Chim. Acta*, *131*, 103 (1981).
35. Guilbault, G. G., Hrabankova, E., *Anal. Chem.*, *42*, 1779 (1970).
36. Guilbault, G. G., Lubrano, G. J., *Anal. Chim. Acta*, *69*, 183 (1974).
37. Ianniello, R. M., Yacynych, A. M., *Anal. Chem.*, *53*, 2090 (1981).
38. Guilbault, G. G., Shu, F. R., *Anal. Chem.*, *44*, 2161 (1972).
39. White, W. C., Guilbault, G. G., *Anal. Chem.*, *50*, 1481 (1978).
40. Fung, K. W., Kuan, S. S., Sung, H. Y., Guilbault, G. G., *Anal. Chem.*, *51*, 2319 (1979).

41. Wawro, R., Rechnitz, G. A., *J. Membr. Sci.*, *1*, 143 (1976).
42. Hsiung, C. P., Kuan, S. S., Guilbault, G. G., *Anal. Chim. Acta*, *90*, 45 (1977).
43. Kovach, P. M., Meyerhoff, M. E., *Anal. Chem.*, *54*, 217 (1982).
44. Toyota, T., Kuan, S. S., Guilbault, G. G., *Anal. Chem.*, *57*, 1925 (1985).
45. Thompson, H., Rechnitz, G. A., *Anal. Chem.*, *46*, 246 (1974).
46. Winquist, F., Lundstrom, I., Danielsson, B., *Anal. Chem.*, *58*, 145 (1986).
47. Tsuchida, T., Yoda, K., *Clin. Chem.*, *29*, 51 (1983).
48. Danielsson, B., Lundström, I., Mosbach, K., Stiblet, L., *Anal. Lett.*, *12*, 1189 (1979).
49. Papariello, G. J., Mukherji, A. K., Shearer, C. M., *Anal. Chem.*, *45*, 790 (1973).
50. Cullen, L. F., Rusling, J. F., Schleifer, A., Papariello, G. J., *Anal. Chem.*, *46*, 1955 (1974).
51. Caras, S., Janata, J., *Anal. Chem.*, *52*, 1935 (1980).
52. Kawashima, T., Rechnitz, G. A., *Anal. Chim. Acta*, *83*, 9 (1976).
53. Nanjo, M., Guilbault, G. G., *Anal. Chem.*, *46*, 1769 (1974).
54. Nabi Rahni, M. A., Guilbault, G. G., de Olivera, N. G., *Anal. Chem.*, *58*, 523 (1986).
55. Hahn, Y., Olson, C. L., *Anal. Chem.*, *51*, 444 (1979).
56. Huang, H. S., Kuan, S. S., Guilbault, G. G., *Clin. Chem.*, *23*, 671 (1977).
57. Kumar, A., Christian, G. D., *Clin. Chim. Acta*, *74*, 101 (1977).
58. Clark, L. C., Duggan, C., Grooms, T. A., Hart, L. M., Moore, M. E., *Clin. Chem.*, *27*, 1978 (1981).
59. Malinauskas, A., Kulys, J., *Anal. Chim. Acta*, *98*, 31 (1978).
60. Engstrom, R. C., Ph.D. Thesis, University of Wisconsin–Madison, 1979.
61. Blaedel, W. J., Wang, J., *Anal. Chem.*, *52*, 1426 (1980).
62. Cheng, F. S., Christian, G. D., *Clin. Chem.*, *24*, 621 (1978).
63. Smith, M. D., Olson, C. L., *Anal. Chem.*, *47*, 1074 (1975).
64. Blaedel, W. J., Jenkins, R. A., *Anal. Chem.*, *48*, 1240 (1976).
65. Williams, D. L., Doig, A. R., Korosi, A., *Anal. Chem.*, *42*, 118 (1970).
66. Saud-Chagny, M. F., Gonon, F. G., *Anal. Chem.*, *58*, 412 (1986).
67. Bartalits, L., Nagy, G., Pungor, E., *Clin. Chem.*, *30*, 1780 (1984).
68. Deng, I., Enke, C., *Anal. Chem.*, *52*, 1937 (1980).
69. McKenna, K., Brajter-Toth, A., *Anal. Chem.*, *59*, 954 (1987).
70. Seegopaul, P., Rechnitz, G. A., *Anal. Chem.*, *56*, 852 (1984).
71. Kihara, K., Yasukawa, E., Hirose, S., *Anal. Chem.*, *56*, 1876 (1984).
72. Fonong, T., Rechnitz, G. A., *Anal. Chim. Acta*, *158*, 357 (1984).
73. Baum, G., *Anal. Lett.*, 39, 65 (1970).
74. Ho, M. Y. K., Rechnitz, G. A., *Anal. Chem.*, *59*, 536 (1987).
75. Arnold, M. A., Rechnitz, G. A., *Anal. Chem.*, *52*, 1170 (1980).
76. Arnold, M. A., Rechnitz, G. A., *Anal. Chem.*, *53*, 515 (1981).
77. Rechnitz, G. A., Arnold, M. A., Meyerhoff, M. E., *Nature* (London), *278*, 466 (1979).
78. Smit, N., Rechnitz, G. A., *Biotech. Lett.*, *6*, 209 (1984).
79. Macholan, L., Chmelikova, B., *Anal. Chim. Acta*, *185*, 187 (1986).
80. Uchiyama, S., Rechnitz, G. A., *Anal. Lett.*, *20*, 451 (1987).
81. Uchiyama, S., Rechnitz, G. A., *J. Electroanal. Chem.*, *222*, 343 (1987).
82. Updike, S., Treichel, I., *Anal. Chem.*, *51*, 1643 (1979).
83. Sidwell, J. S., Rechnitz, G. A., *Biotech. Lett.*, *7*, 419 (1985).
84. Wang, J., Lin, M. S., *Anal. Chem.*, in press (7/1988).
85. Arnold, M. A., Rechnitz, G. A., *Anal. Chem.*, *53*, 1837 (1981).
86. Arnold, M. A., Rechnitz, G. A., *Anal. Chem.*, *54*, 777 (1982).
87. Rechnitz, G. A., Kobos, R. K., Richel, S. J., Gebauer, C. R., *Anal. Chim. Acta*, *94*, 357 (1977).
88. Karube, I., Okada, T., Suzuki, S., *Anal. Chem.*, *53*, 1852 (1981).
89. Karube, I., Nakahara, T., Matsunaga, T., Suzuki, S., *Anal. Chem.*, *54*, 1725 (1982).
90. Di Paolantonio, C. L., Rechnitz, G. A., *Anal. Chim. Acta*, *141*, 1 (1982).

91. Di Paolantonio, C. L., Rechnitz, G. A., *Anal. Chim. Acta, 148,* 1 (1983).
92. Matsumoto, K., Seijo, H., Watanabe, T., Karube, I., Satoh, I., Suzuki, S., *Anal. Chim. Acta, 105,* 429 (1979).
93. Simpson, D. L., Kobos, R. K., *Anal. Lett., 15,* 1345 (1982).
94. Kobos, R. K., *Trends Anal. Chem., 6,* 6 (1987).
95. Aizawa, M., Morioka, A., Suzuki, S., *Anal. Chim. Acta, 115,* 61 (1980).
96. Di Gleria, K., Hill, H. A. O., McNeil, C. J., Green, M. J., *Anal. Chem., 58,* 1203 (1986).
97. Keating, M. Y., Rechnitz, G. A., *Anal. Chem., 56,* 801 (1984).
98. Shiba, K., Umezawa, Y., Watanabe, T., Ogawa, S., Fujiwara, S., *Anal. Chem., 52,* 1610 (1980).
99. Haga, M., Sugawara, S., Itagaki, H., *Anal. Biochem., 118,* 286 (1981).
100. D'Orazio, P., Rechnitz, G. A., *Anal. Chim. Acta, 109,* 25 (1979).
101. Krull, U. J., *Anal. Chim. Acta, 192,* 321 (1987).
102. Krull, U. J., Thompson, M., *Trends Anal. Chem., 4,* 90 (1985).
103. Yao, T., Rechnitz, G. A., *Anal. Chem., 59,* 2115 (1987).
104. Ikariyama, Y., Furuki, M., Aizawa, M., *Anal. Chem., 57,* 496 (1985).
105. Rechnitz, G. A., *Science, 214,* 287 (1981).
106. Arnold, M. A., *Am. Lab., 15*(6), 34, (1983).
107. Rechnitz, G. A., *Anal. Chim. Acta, 180,* 281 (1986).
108. Borman, S., *Anal. Chem., 59,* 1091 (1987).
109. Turner, P. F., Karube, I., Wilson, G. S., *Biosensors,* Oxford Science Publications, Oxford, 1987.

REFERENCES

C H A P T E R 4

Electrochemical Detection for Liquid Chromatography and Automated Flow Systems

4-1 LIQUID CHROMATOGRAPHY WITH ELECTROCHEMICAL DETECTION

The coupling of liquid chromatography with electrochemistry (LCEC) offers a selective and sensitive tool for the determination of a wide variety of compounds in body fluids and tissues. In addition to its inherent sensitivity and selectivity, electrochemical detection possesses the important features of low dead volume, fast and linear response, and low cost. Over the past 15 years, LCEC has evolved and grown into a powerful and versatile analytical technique. Nowadays, LCEC is commonplace in many clinical, biochemical, and pharmaceutical laboratories. In addition, electrochemical detectors are often coupled with other flow systems, e.g., segmented and nonsegmented flow automatic analyzers, that are common in the clinical laboratory. The principles of several electroanalytical techniques, e.g., voltammetry, potentiometry, and conductometry, have been used in conjunction with electrochemical detectors for liquid chromatography and automated flow systems.

In the following sections the principles, requirements, and clinical and biomedical applications of LCEC will be described.

4-1.1 Voltammetric Detection

4-1.1.1 PRINCIPLES

Voltammetric detectors can be used to detect compounds that are electroactive. This is both an advantage and a limitation: an advantage because the detector can be largely selective, and a limitation since only electroactive compounds are detectable. Electrochemical detection is usually performed by controlling the potential of the working electrode at a fixed value (corresponding to the limiting current plateau region of the compounds of interest) and monitoring the current as a function of the elution time. The current peaks generated represent the concentration profiles of the eluting compounds as they pass through the detector.

109

Accordingly, the magnitude of the peak current serves as a measure of the concentration. The current peaks are superimposed on a constant background current (caused by redox reactions of the mobile phase). Larger background currents, expected at high potentials, result in increased (flow-rate-dependent) noise level. Increased noise level, associated with the presence of dissolved oxygen in the mobile phase, hampers the cathodic detection of reducible species. The signal-to-noise ratio provides a measure of the minimum detectable quantity of the analyte. Since each class of compounds exhibits different electrochemical behavior (Figure 4-1), the operating potential can be exploited for selective detection (see the discussion below). In general, a lower potential is more selective, and a higher potential more universal. Thus, compounds undergoing redox reactions at lower potentials can be determined with greater selectivity.

Proper use of LCEC requires knowledge of the redox reactions and their dependence on the composition of the mobile phase. In most cases, a short cyclic voltammetry experiment can provide the desired information. This type of experiment is performed in a batch cell containing a quiescent solution that resembles the one found in the LCEC detector. A more accurate (but time-consuming) approach would be the construction of hydrodynamic voltammograms. This approach commonly serves for selecting the applied potential for the amperometric detection. Hydrodynamic voltammograms can be obtained by making

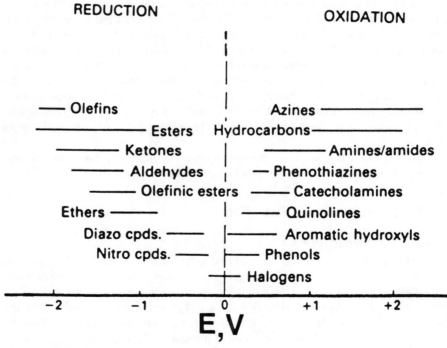

FIGURE 4-1
Electroactivity of a range of organic functional groups.

repeated injections of the analyte solution while recording the current at different potentials. (The easiest way to accomplish this is by the flow injection method.) The resulting voltammograms have a wave (sigmoidal) shape, characterized by a half-wave potential and a limiting current region. Figure 4-2 illustrates hydrodynamic voltammograms for several oxidizable compounds, A, B, and C. Although it is common to operate the detector on the limiting current plateau region, a lowering of the operating potential (to the rising portion of the wave) can be used to improve the selectivity and lower the detection limit. For example, if compounds B and C coelute, lowering the potential to E_1 will allow selective monitoring of B. This procedure has been used to measure trace quantities of p-aminophenol in the drug acetaminophen.[1] Comparison of hydrodynamic voltammograms for the sample peak and a standard can provide important information regarding the peak identity. The utility of this approach in measurements of serotonin in urine is illustrated in Figure 4-3.

4-1.1.2 MOBILE PHASE AND SAMPLE REQUIREMENTS

As in other voltammetric measurements, a conductive solvent is needed to maintain good potential control. Because of the polar nature of their mobile phase, reversed-phase and ion-exchange separations offer better compatibility with electrochemical detection. The unsuitability of electrochemical detection for normal-phase chromatography has been attacked using various approaches. These include a postcolumn addition of an electrolyte solution,[3] use of a large-volume wall-jet detector,[4] and ultramicro working electrodes.[5] It is often important to buffer the pH of the mobile phase because most redox reactions of biological compounds either release or consume protons. Electrolyte or buffer concentrations ranging from 0.01 to 0.1 M are sufficient. As the mobile-phase composition or pH is important for both the chromatographic separation and electrochemical detection, a compromise may be required in certain situations. The purity of the solvents and

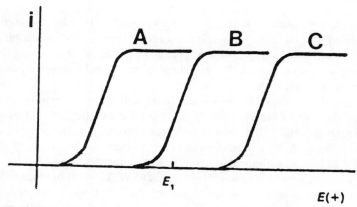

FIGURE 4-2
Hydrodynamic voltammograms for three oxidizable compounds.

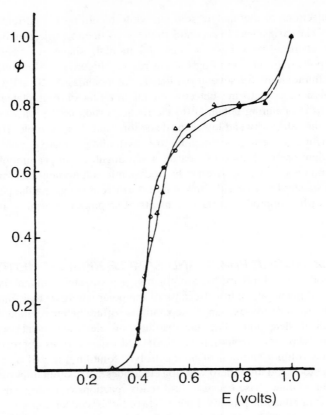

FIGURE 4-3
Hydrodynamic voltammograms of serotonin standard (o) and the amberlite eluate from urine specimen (Δ). φ is the normalized current function. (Reproduced with permission.[2])

electrolyte used is critical to minimize the background current and the corresponding noise level. For the same reason, it is necessary to purge the mobile phase with inert gas—when monitoring reducible species—in order to remove the electroreducible oxygen. (Stainless-steel connecting tubing should be used to prevent reaeration of the mobile phase.)

Most LCEC applications involving biological samples require a cleanup procedure. The type of sample cleanup necessary for a particular analysis depends on the biological matrix, the selectivity of the chromatographic system, and the required sensitivity. Such a cleanup procedure minimizes fouling of the electrode by proteins or lipids, and eliminates many potential interferences. This is usually accomplished by removal of proteins by precipitation, followed by extraction of the electroactive species of interest into a suitable solvent. Some samples, such as urine, can be injected following dilution (e.g., 1:40) with the mobile phase and filtration to remove colloidal materials. The use of a precolumn is recommended in the latter case to prolong the life of the analytical column.

4-1.1.3 ELECTRODES

The choice of the working electrode material for LCEC depends mainly on two factors: the redox behavior of the analytes of interest and the background current at the potential used for detection. The most commonly used electrode materials in LCEC are glassy carbon, carbon paste, mercury, and platinum.

Carbon-based electrodes have typically served as working electrodes of choice for detection of oxidizable species in the anodic region. The most versatile choice is glassy carbon; it is inert to common organic solvents, is impermeable to gas, and has a wide potential range. Before operation, glassy carbon should be carefully polished to a mirrorlike appearance using standard metallographic procedures. Carbon paste electrodes, which use graphite powder mixed with organic binder (e.g., oil, grease), offer the advantages of low background current and noise levels, low cost, and an easily renewable surface. The most severe limitation of carbon paste electrodes is the tendency of the organic binder to dissolve in mobile phases containing an appreciable fraction (20–30%) of organic solvent. Another carbon-based composite electrode material is the Kelgraf electrode, made from a mixture of graphite and Kel-F.[6] This material is compatible with a wide range of organic solvents. The unique mass-transport properties of composite (microarray) electrodes result in improved signal-to-noise characteristics as compared with continuous electrodes. Among the various forms of porous carbon flow-through electrodes, reticulated vitreous carbon (RVC) appears to be especially suitable for LCEC operation.[7, 8] Operation of carbon-based electrodes at high sensitivity requires 30–60 min of daily start-up to achieve stable baselines after applying the operation potential. This is due to the slow equilibration of the electrode surface at any given new potential.

The analytical capability of carbon electrodes can be enhanced using various surface modification and electrochemical pretreatment procedures. For example, a simple electrochemical pretreatment of glassy carbon electrodes has been shown to increase the selectivity of LCEC detection by lowering the required operating potential.[9, 10] As a result, quantitation of hydrazines, ascorbic acid, and NADH in urine samples can be significantly simplified. Similarly, coating a glassy carbon electrode with cellulose acetate film excludes macromolecules from the surface.[11, 12] This results in improved specificity toward smaller analytes and minimization of electrode poisoning due to protein adsorption. Similarly, detectors coated with charged polymeric films, e.g., poly(4-vinylpyridine) or nafion can be employed to improve the specificity toward counterionic analytes.[13]

Applications of platinum and gold electrodes have found increasing use for triple-pulse amperometric detection of compounds, e.g., alcohols, amino acids, or carbohydrates, undergoing surface-catalyzed oxidation.[14, 15] Such compounds may also be detected using nickel and copper working electrodes.[16, 17] The investigation of new (or improved) electrode materials will continue to receive a great deal of attention in the near future. Modified electrodes will play an important role in these studies.

4-1.1.4 DETECTORS

A variety of voltammetric detectors has been described in the literature, and several are available from commercial sources (e.g., Bioanalytical Systems, waters, Metrohm, Environmental Sciences Associates, Tacussel, EG&G Princeton Applied Research, Dionex). The cell design must fulfill certain requirements, including high sensitivity and precision, minimal dead volume, and ease of construction and maintenance. Generally, flow-through voltammetric detectors can be divided into two categories: those that electrolyze only a small fraction of the analyte, typically 1–10% (amperometric detectors), and those for which the electrolytic efficiency is 100% (coulometric detectors). The methodology of applying the potential and measuring the current response is identical for both categories. (In that sense, the so-called coulometric detectors are also amperometric.) A body of theory has been developed to allow optimization of the detector design and performance. As shown in Figure 4-4, basically three designs have been used as low-yield amperometric detectors: thin-layer, wall-jet, and tubular configurations. In the commonly used thin-layer configuration, a thin layer of solution flows parallel to the electrode surface in a rectangular channel. In the wall-jet design, the stream flows from a nozzle perpendicularly onto the electrode surface. Working electrodes with a diameter of 2–4 mm are used in these configurations, with the reference and counter electrodes being located downstream. A large-volume wall-jet cell—with larger distance between the solution inlet and the working electrode—extends the scope of LCEC to normal-phase separations.[4] The tubular design employs an open tube as the working electrode, through which the solution flows. The limiting current response i_1 of these configurations is described by the general equation

$$i_1 = nFAKCU^\alpha \tag{4-1}$$

where n is the number of electrons, F is the value of Faraday, A is the electrode area, C is the bulk concentration of the electroactive species, U is the flow rate, and K and α are constants (dependent on the flow regime and geometry). The experimental peak current differs from this theoretical steady-state dependence, as it is influenced by the chromatographic band broadening.[19]

Coulometric detectors require electrodes with a large surface area to achieve complete electrolysis. Various porous electrode materials have been used for this purpose. Under these conditions, the limiting current is given by Faraday's law:

$$i = nFCU \tag{4-2}$$

The increased response associated with these detectors is coupled with larger noise and background current. Accordingly, coulometric (high-conversion efficiency) detection has no advantage over amperometric (low-conversion efficiency) detection in terms of detection limits.

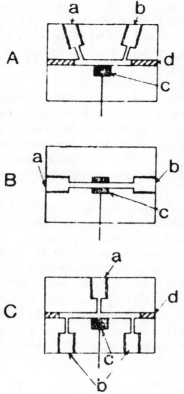

FIGURE 4-4
Diagrams of common electrochemical detectors: (A) thin-layer channel; (B) tubular; (C) wall jet; (a) inlet; (b) outlet; (c) working electrode; (d) spacer. (Reproduced with permission.[18])

Dual-electrode LCEC detectors, both amperometric and coulometric, are available from various sources. Such detectors offer various advantages, as described in Section 4-1.1.5.2.

Various flow cells with a dropping mercury electrode have been described.[20] The major problems with these cells are the difficulties in attaining low dead volume and reproducible drop size. These problems have been solved in the EG&G commercial detector, where the mercury drop is combined with the wall-jet principle.[21] The effective dead volume of this detector is about 1 μl.

Various submicroliter-volume electrochemical detectors have been examined for use with microbore liquid chromatography.[22, 23] The nature of electrochemical processes, which occur at a surface rather than in volume, makes these attractive for miniaturized detection. Unlike other detection schemes, the sensitivity of the electrochemical detection is not sacrificed by the reduction in dead volume. Such operation allows, for example, measurements of urinary metabolites in urine samples as small as 0.1 μl.[24] In addition, an on-column electrochemical detector for open-tubular liquid chromatography, using a single carbon fiber, has been applied successfully for the detection of electroactive compounds in urine.[25] Microelectrodes also offer improved signal-to-noise characteristics. Several review articles describe the different types of electrochemical detectors.[18, 20, 26]

4-1.1.5 DETECTION MODES

The simplest and by far the most common detection scheme is the measurement of the current at a constant potential. Such constant-potential measurements have the advantage of being free of double-layer charging and surface transient effects. As a result, extremely low detection limits — on the order of 1 – 100 pg (about 10^{-14} moles of analyte)—can be achieved. In various situations, however, it may be desirable to change (pulse, scan, etc.) the potential during the analysis.

4-1.1.5.1 *Potential Pulse*.   Several potential-pulse detection schemes have been applied to LCEC. Differential-pulse amperometry has been shown to improve the selectivity for compounds that react at potentials higher than other coeluting compounds.[27] For this purpose the potential of the working electrode is pulsed around a base potential, chosen around the half-wave potential of the analyte, and the difference in the current before and at the end of the pulse is measured (Figure 4-5). This operation is responsive only to compounds with redox potentials within the pulse width. Reversed-pulse amperometry, in which the current is sampled only once, at the end of the pulse, appears to have great potential for minimizing the oxygen interference when reducible species are monitored.[28] Such pulse techniques

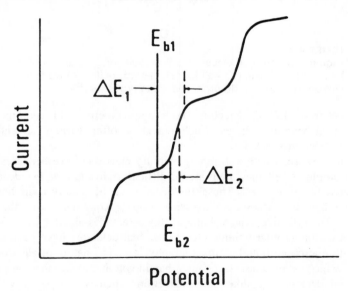

FIGURE 4-5
Differential-pulse detection. ΔE is the pulse height; E_b is the base potential. (Reproduced with permission.[27])

have larger background current and noise level than constant potential measurements, especially when carbon electrodes are used.

Johnson and coworkers [14, 15] developed a triple-pulse waveform that allows effective detection of organic compounds such as carbohydrates, alcohols, or amino acids, which undergo surface-catalyzed oxidation at platinum or other noble electrodes. Such compounds cannot be detected by constant potential measurements due to the irreversible poisoning of the electrode surface. The triple-pulse waveform incorporates the amperometric measurement with a potentiostatic cleaning and reactivation of the surface within a single waveform (Figure 4-6). Because the mechanism of the detection of alcohols and carbohydrates involves prior adsorption, the shape of the calibration plot reflects the corresponding adsorption isotherm. The incorporation of triple-pulse amperometry with ion chromatography systems offers a powerful tool, especially when CHOH-bearing compounds are measured. This concept of detection is now being commercialized by the Dionex Corporation.[29]

It is possible also to apply potential scanning techniques for improving the selectivity of LCEC detection. Complete voltammograms, recorded at short time intervals during the elution, can give virtually instantaneous electroanalytical data. The addition of the redox potential selectivity of voltammetry helps to identify a component and to resolve components not completely resolved by the column. For example, potential scanning detection can be used to resolve the coeluting peaks of L-dopa, 6-hydroxydopa, and tyrosine.[30] Different approaches to swept-potential detectors based on square-wave voltammetry,[31-33] phase-sensitive ac voltammetry,[34] normal-pulse voltammetry,[35] or coulostatic measurements[36, 37] have been reported. Detection limits of these detectors are usually 1 to 2 orders of magnitude

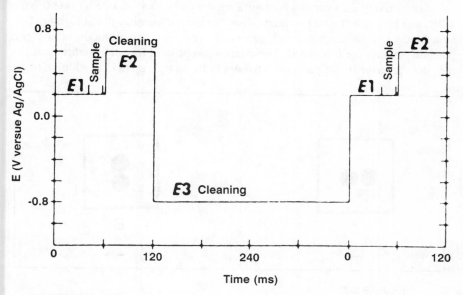

FIGURE 4-6
Triple-pulse amperometric waveform.

higher than those obtained with amperometric detection (owing to the additional background current associated with the potential scan).

4-1.1.5.2 Dual-Electrode Detection.

Recent work has shown that both the selectivity and sensitivity of electrochemical detectors can be improved by using more than one working electrode.[38, 39] Various important clinical applications of dual-electrode cells have been reported, using different orientations of the two working electrodes with respect to each other and the flow axis (Figure 4-7). For example, in some situations it is advantageous to convert the analyte from one chemical form to another, for more selective and sensitive detection. For this purpose, an upstream working electrode generates products that can be detected at the downstream electrode held at a more favorable potential (Figure 4-7A). This series configuration can be useful for detection of reducible compounds, which suffers from large background current when using a single electrode. For example, Allison and Shoup[40] described the simultaneous detection of disulfides and thiols using a mercury dual-electrode thin-layer detector. In this case, the upstream electrode—kept at an extreme negative potential—served to convert the disulfides to the parent thiols. Both the converted and original thiols were detected at the downstream electrode. The method was applied to the determination of oxidizable and reducible glutathione in human blood. The detector selectivity can also be improved using the series configuration, because many electroactive compounds do not produce detectable products. The series dual-electrode approach is not limited to the thin-layer amperometric detector; coulometric electrodes have been shown to give similar advantages.[39]

Alternatively, in a parallel configuration the electrodes are side by side (Figure 4-7B) and held at different potentials along the hydrodynamic voltammogram. Two simultaneous chromatograms are generated, and the current ratios at the two potential settings are calculated. These ratios can be used for peak confirmation, as they are characteristic of particular analytes in the sample matrix. In addition to the

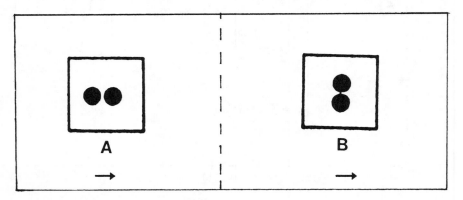

FIGURE 4-7
Dual-electrode configurations for operation in the series (A) and parallel (B) modes.

improvement in qualitative information, the parallel dual-electrode arrangement can be useful when one wishes to quantitate easily oxidizable (or reducible) compounds in the presence of others that react at higher potentials. For example, Figure 4-8 shows chromatograms of a urine sample obtained 4 h after a normal dosage of acetaminophen (two Tylenol tablets). Most of the major peaks are acetaminophen metabolites. The more thoroughly metabolized compounds have lower oxidation potentials, and thus can be detected more favorably at the lower potential electrode. Similar improvement in the selectivity and peak identity assignments were obtained in the determination of pterins in biological samples.[41] Other clinical applications of dual-electrode LCEC systems are described in Section 4-1.3. The same dual-electrode cell, with different orientations of the electrode, can be used to obtain the series and parallel configurations.

In conclusion, no single operation mode, cell configuration, or electrode material satisfies in all situations. Proper choice of these would depend upon the specific application. Such biomedical and clinical applications of LCEC are reviewed in the following sections. The reader is referred to several review articles[20, 26, 42, 43] for additional information on the design and operation of voltammetric detectors.

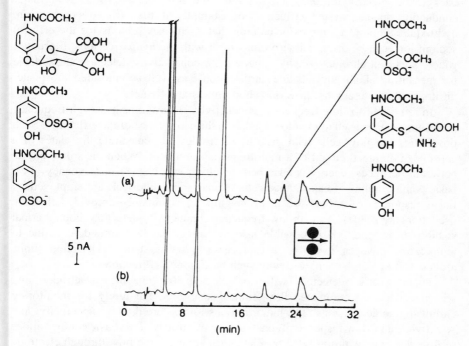

FIGURE 4-8
Chromatograms of a urine sample obtained 4 h after normal acetaminophen dosage. Simultaneous parallel-adjacent detection with electrodes at (a) +1.0 and (b) +0.65 V. (Reproduced with permission.[38])

4-1.2   Conductivity and Potentiometric Detections

In addition to voltammetric detectors that are commonly used for trace organic analysis, conductivity and potentiometric detectors are often used to monitor ionic species in chromatographic effluents and other flowing streams.

Conductivity detectors are sensitive to ionic solutes in a medium of low conductivity. Such detectors offer simple design and operation, wide linear dynamic range (up to six orders of magnitude[44]), universal response to ionic species, and small dead volumes [down to 0.5 μl (Ref. 45)]. The cell consists of two electrodes, usually made of platinum, through which the column effluent flows. A constant alternating potential is applied to the electrode, and the resulting current is monitored. This current provides a measure of the solution conductivity. The temperature must be controlled carefully, because the response is temperature-dependent.

Using a mobile phase with high conductivity decreases the detector sensitivity. The technique of background conductivity suppression is usually used to cope with the problem of the effluent background.[46] It involves the use of a second (suppressor) column, downstream from the analytical column. This removes the background electrolyte ions, leaving only the ions of interest as the major conducting species in the effluent. For example, if the effluent is a sodium hydroxide solution, a suppressor column that exchanges protons for the effluent sodium ions can be used. These protons react with the hydroxide ions to produce water. Such ion chromatography systems with conductivity detection are suitable for measuring alkali and alkaline earth cations, as well as chloride, phosphate, nitrite, and sulfate anions, in various body fluids and tissues.[47]

In addition to low-frequency conductivity measurements, high-frequency procedures can be used to monitor changes in the capacitance of the effluent.[48] Such procedures use flow cells with metallic electrodes that constitute the plates of a capacitor in a tuned circuit of a radiofrequency oscillator. When the solute passes between the plates, energy is adsorbed, resulting in a change of frequency. Two basic configurations of capacitance cells, planar and cylindrical, are employed. A linear response, over four orders of magnitude, is obtained (provided that the oscillators are stable). As with low-frequency conductivity detectors, good thermal stability is required. It is possible also to correct the background drift due to temperature changes using a microprocessor-based system.[49] Detection limits around 150 ng can be achieved with such background correction.

Potentiometric detectors with ion-selective (or enzyme) electrodes find widespread use in automated flow clinical analyzers, but rarely for monitoring chromatographic effluents. The major reasons for this are the high specificity, low sensitivity, and slow response of ion-selective electrodes. Two basic configurations of flow detectors with ion-selective electrodes are used: the flow-through electrode (such as tubular steady electrodes[50]) and the cap design.[51] The latter consists simply of an electrode probe fitted tightly with a cap, with an inlet and outlet for the flowing stream (Figure 4-9). Both configurations can be built with volumes of about 5 μl.

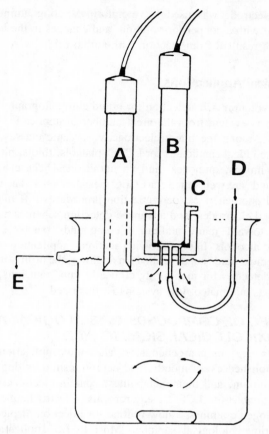

FIGURE 4-9
Flow-through cell—cap design. (A) reference electrode, (B) ion-selective electrode, (C) flow-through cap, (D) inlet, (E) waste. (Reproduced with permission.[51])

The reference electrode is placed downstream from the indicator electrode, so that leakage of ions from the former does not affect the response. Coated-wire electrodes and field effect transistors, in which the internal reference solution is eliminated, are also well suited for low-volume potentiometric flow cells.

The detector response is related to the concentration of the ion of interest in the flowing stream via the Nernst equation (Equation 1-2). Detection limits are usually in the 10^{-5}–$10^{-6}M$ range. The response time of potentiometric detectors can be improved, e.g., by sweeping the surface with a bubble, to yield high-speed automated flow systems.[52] The high specificity "problem," which precludes wide use of these detectors in liquid chromatography, can be attacked by using multielectrode detection, by using electrodes of poor selectivity, or by titrating the eluting components with an ion sensed by the detector. For example, a copper-selective electrode was used for monitoring eluting amino acids.[53] The effluent was mixed with a copper ion solution, and changes in the level of the free copper ions, which resulted from reaction with amino acids, were measured.

4-1.3   Clinical Applications

The popularity of voltammetric detection for liquid chromatography for analysis of biological samples is evident from the great number of applications reported over the last 15 years. Among the many electroactive organic classes, the following compounds are often ideal candidates for LCEC: phenols, thiols, nitro compounds, aromatic amines, imines, quinones, and phenothiazines. These and many other biological compounds are well suited for LCEC measurements. The development of new chemical and photochemical derivatization procedures (off-line and on-line), chemically modified electrodes, and advanced detection schemes has extended the scope of LCEC toward many additional compounds, such as amino acids, carbohydrates, and alcohols. In the following sections, applications of LCEC in the clinical and biomedical fields are described. In addition, the utility of potentiometric and conductivity detectors for monitoring clinically important ions, enzymes, and substrates in various automated flow systems is illustrated.

4-1.3.1 LCEC OF COMPOUNDS OF BIOLOGICAL AND TOXICOLOGICAL SIGNIFICANCE

LCEC got its start in neurochemistry, and new applications in this field continue to be published every month. The neurotransmitters dopamine, norepinephrine, and serotonin, and their metabolites, contain electroactive phenolic or indole moieties suitable for LCEC measurements. Electrochemical detection of these and other biogenic amines is now routine for injection of picomole amounts isolated from various biological samples. Most LCEC applications have been concerned with the determination of neurologically important biogenic amines in brain tissues or body fluids following specific treatment (e. g., drug administration) or in disease models. Characteristic data resulting from such experiments are illustrated in Figure 4-10. Liquid-solid extraction of the neurotransmitters onto alumina, following by their elution with dilute acids, is usually employed for isolating these compounds from complex biological matrices. Brain studies can be substantially improved by dialyzing the extracellular fluid prior to exit from the tissue. Hence, the cleanup procedure is eliminated, and perturbation of the brain environment is minimized. In particular, on-line microdialysis/small-bore liquid chromatography has been extremely useful for the rapid (5-min) analysis of extracellular dopamine.[55] Urine generally requires an additional cleanup step. Representative papers, describing these applications, are given below.

Riggin and Kissinger[56] reported a sensitive, specific, and rapid procedure for the determination of catecholamines in urine by reversed-phase liquid chromatography with electrochemical detection. Mapping the levels of neurochemically important compounds in brain tissue is one of the most important contributions of LCEC. For example, Adams and coworkers[57] studied the lateral distribution of

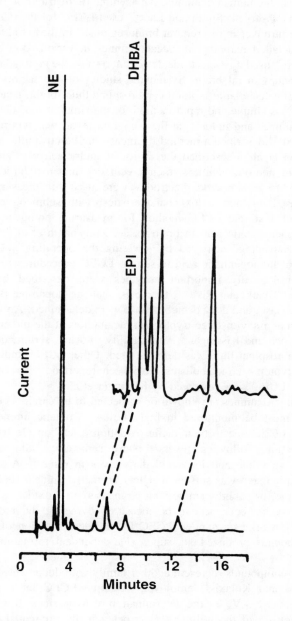

FIGURE 4-10
Chromatograms obtained from 230 mg of heart tissue assayed
using liquid-solid extraction to isolate the cathecols. The sample
contained 165 ng/g norepinephrine, 14 ng/g epinephrine, and
20 ng/g dopamine. (Reproduced with permission.[54])

norepinephrine in the human thalamus. An asymmetric distribution (with left-right differences) was observed. Smith and Lane[58] used LCEC for the determination of biogenic amine turnover in discrete rat brain regions. Mefford et al.[59] used LCEC for obtaining detailed mapping of catecholamines in various dog brain regions. Yamamoto et al.[60] used dual-electrode detection to study the regional distribution of guanine nucleotides in rat brain. Systematic studies on the factors affecting the determination of catecholamines in rat cerebrospinal fluid, brain, heart, and plasma were reported.[61] A simple and rapid method for the simultaneous determination of brain catecholamines and indoles, at the picogram level, was reported by Kim et al.[62] Harris et al.[63] described a method for measuring homovanillic acid in human plasma. Matson et al.[64] described the utility of multielectrode detectors for the determination of neurotransmitters, their precursors, and metabolites, in various tissues. Up to 15 coulometric electrodes were used for measuring coeluting compounds based on their redox characteristics, with subpicogram sensitivity. Hefti[65] reported a simple LCEC method for measuring homovanillic acid and 3,4-dihydroxyphenylacetic acid in brain tissue. Zoutendam et al.[66] demonstrated the selectivity advantage achieved by lowering the operating potential for the determination of homogentistic acid in serum. LCEC procedures for the determination of neurologically important enzymes were described by Davis and Kissinger[67] and Blank and Pike.[68] Enzymes such as dopamine-β-hydroxylase, tyrosine hydroxylase, and dopa decarboxylase were determined. An LCEC method for the concurrent assay of three tryptophan metabolites at the picomole level was described by Koch and Kissinger.[2] Applicability to urine, serum, tissue homogenates, and cerebrospinal fluid was demonstrated. Other LCEC studies concerning tyrosine and tryptophan metabolisms have been reported.[69] The neurochemical applications of LCEC were reviewed by Kissinger et al.[70]

Many aromatic amines are known or suspected to be carcinogenic in humans and therefore must be monitored in body fluids. Aromatic amines are readily determined by LCEC due to their facile oxidation at carbon electrodes. Electrochemical detection, following reversed-phase separation, allowed convenient monitoring of benzidine and its acetyl derivatives in urine.[71] A simple solvent extraction isolation step was sufficient. Other aromatic amines, such as aminoanthracene, tolidine, or naphthylamine, can be measured by oxidative LCEC, while nitrobiphenyl and azobenzene can be measured by using the reductive mode. Kissinger and coworkers[72] applied LCEC for the determination of hydroxylated aromatic compounds produced via superoxide-dependent formation of hydroxyl radicals.

Hydrazine compounds represent another family with documented carcinogenic behavior. Fiala and Kulakis[73] employed oxidative LCEC, at a glassy carbon electrode held at +1.0 V, for the determination of hydrazines. Ravichandran and Baldwin[9] demonstrated the utility of electrochemically pretreated glassy carbon electrodes for the direct measurement of simple hydrazines in urine.

Many sulfur-containing biological compounds are good candidates for LCEC. LCEC procedures for thiols are based on the unique reaction between these compounds and the mercury surface, which involves formation of a stable complex.

The resulting current is actually due to the oxidation of mercury. Such LCEC measurements of thiols were first proposed by Rabenstein and Saetre,[74, 75] who used a mercury pool working electrode. Selective measurements of glutathione in blood were reported. The use of dual mercury electrodes in the series configuration for the simultaneous determination of thiols and disulfides was reported by Allison and Shoup.[40] This procedure was also applied for the determination of glutathione in whole blood samples. Stein et al.[76] described a similar procedure for trace measurements of thiols and disulfides in liver and kidney tissues; typical chromatograms are shown in Figure 4-11. Shaw et al.[77] described the measurement of low concentrations of penicillamine in plasma by LCEC with a glassy carbon electrode held at +0.8 V. Carbon paste electrodes modified with cobalt phthalocyanine were used for amperometric detection of glutathione and cysteine in plasma and blood following minimal sample preparation.[78] The electrocatalytic nature of the electrode enabled lower potential operation, and therefore enhanced selectivity, compared with unmodified electrodes.

The pterins are a family of heterocycles that are cofactors to several hydroxylases. Lunte and Kissinger[41] described an LCEC procedure for the determination of pterins in biological samples. The dual-electrode approach, in the parallel configuration, was used for simultaneous determination of several pterins occurring in different oxidation states. The same authors reported an investigation of the tautomerization of quinonoid dihydropterins by LCEC.[79]

Uric acid is a major constituent of body fluids, the levels of which may be related to various disease states. Simple and rapid LCEC procedures for measuring uric acid in biological samples, using ion-exchange or reversed-phase columns, have been reported.[80–82] Simultaneous determination of uric acid, allopurinal, and oxipurinol in human serum and urine was described by Palmisano et al.[83] Kafil et al.[84] described the quantitation of nucleic acids at the picogram level using high-performance liquid chromatography with amperometric detection. The method is based on hydrolysis and quantitation of adenine (detection limit = 0.1 pmol) and guanine (detection limit = 0.05 pmol). Stulik and Pacakova[85] described the liquid chromatographic separation of biologically important pyrimidine derivatives, using a carbon paste detector (for monitoring oxidizable derivatives) and a mercury electrode flow cell (for measuring reducible ones). The determination of oxalic acid in biological samples was described by Kok et al.[86] Oxalic acid was separated from interfering compounds by using an ion-exchange column, and was detected at a copper working electrode. Mayer and Greenberg[87] used different electrochemical detection modes for monitoring oxalic acid after ion-exchange separation. The same group applied these methodologies for the determination of oxalic acid in urine[88]

The concentrations of thyroid hormones in plasma can be indicative of a malfunctioning thyroid gland. Purdy and coworkers[89] described a sensitive oxidative LCEC method for an assay for the thyroid hormones. The method permits simultaneous measurement of total serum thyroid hormones, with detection limits at the subnanogram level. The electroactivity of estrogenic steroid hormones makes them good candidates for LCEC detection. Simple LCEC methods for the determination of urinary estriol were reported.[90, 91] Simple solvent extraction

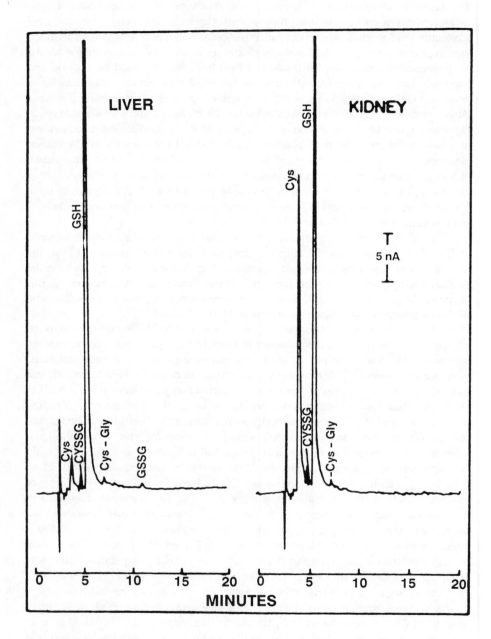

FIGURE 4-11
LCEC response for thiols and disulfides in liver and kidney.
Supernatants from tissues were diluted 1:10 prior to injection.
(Reproduced with permission.[76])

procedures were used, and the minimum detectable concentration was 0.4 μg/ml. Other estrogens, such as estradiol and estrone, can be measured in a similar way. Gunasingham et al.[92] described a method for the determination of oestriol in pregnancy by using normal-phase liquid chromatography and a large-volume wall-jet detector. Oxidative LCEC was applied for measuring diethylstilbestrol in biological samples.[93] Smyth et al.[94] reported an oxidative LCEC procedure for measuring trace levels of phenolic growth-promoting hormones. The steroid hormone lynestrenol was measured by liquid chromatography with tensammetric detection.[95] Kemula and coworkers[96] used reductive LCEC with a rapidly dropping mercury electrode for measurement of testosteroids. The application of LCEC for measurement of 17-ketosteroids in human blood, based on selective derivatization of these compounds, was reported by Shimada et al.[97] Detection limits of 200 pg were reported.

The low redox potential of ascorbic acid makes this vitamin well suited for LCEC measurements. The determination of ascorbic acid in human plasma and urine by LCEC was reported by Mason et al.[98] A simple protein precipitation and dilution with the mobile phase preceded chromatography. Urine samples can be simply filtered and diluted prior to injection.[99, 100] Removal of oxygen from the system is recommended to minimize stability problems. Also, all solutions should be made up (or diluted) in cold dilute perchloric acid, a medium known to inhibit oxidation of ascorbic acid. A procedure for measuring extracellular ascorbic acid in brain was reported by Justice and coworkers.[101] A push-pull cannula perfusion technique was used for sample collection; the mobile phase was 0.1 M sodium acetate (pH 5), containing 1 mM n-octylamine. The LCEC determination of ascorbic acid in complex biological samples was described by Carr and Neff.[102] Brunt and coworkers[103] described the suitability of the dual-electrode LCEC approach for trace measurements of ascorbic acid. LCEC was used also to measure electroactive derivatives and metabolites of ascorbic acid.[104] A series dual-electrode detection system was applied to measure vitamin K in plasma and in rat liver.[105, 106] Vitamin K hydroquinone, generated at the upstream electrode, was detected at the downstream electrode held at a positive potential. Detection limits around 100 pg were reported. Chou et al.[107] described an LCEC method for the determination of vitamin E and its derivatives in serum. The authors reported prior extraction into heptane, and use of δ-tocopherol as an internal standard. LCEC was used also for the determination of naturally occurring vitamin E analogs in human plasma.[108] Huang et al.[109] described the coupling of electrochemical and ultraviolet detections for a simultaneous sensitive analysis of plasma vitamin E and vitamin A. Allenmark et al.[110] described an LCEC procedure for measuring pyridoxal 5′-phosphate in biological materials.

The detection of amino acids at a platinum-wire electrode, operated in the triple-pulse mode, was reported by Polta and Johnson.[111] Both primary and secondary amino acids were monitored in a sodium hydroxide effluent following anion-exchange separation. Allison et al.[112] described the amperometric detection of amino acids, using o-phthalaldehyde as a derivatizing agent. Twenty-two amino acids were separated within 10 min, with detection limits as low as 30 fmol. The use of N,N-dimethylaminophenylisothiocyanate as an electrochemical label for LCEC measurements of amino acids was described by Mahachi et al.[113] The above isothiocyanate reacts with primary and secondary amines to yield easily oxidized phenylthiohydantoins. Wightman and coworkers[114] described another derivatization method of amino acids using dinitrobenzenesulfonyl chloride. LCEC measurements of tyrosine-related peptides were reported by White.[115] Meek [116] described a novel derivatization for the reductive electrochemical detection of peptides at the picomole level, using 3,6-dinitrophthalic anhydride. In addition to these amperometric detection schemes, potentiometric detection—with a copper-selective electrode—can be used for selective measurements of copper-binding amino acids.[117] The selectivity of this method was demonstrated by the analysis of urine with no sample pretreatment other than filtration. LCEC procedures for the determination of neurologically important amino acids (e.g., tyrosine, tryptophan) have been described early in this section.

Triple-pulse amperometry at a platinum wire electrode was demonstrated as an effective detection scheme for carbohydrates following chromatographic separation[14]; the detection limits were 100–500 ng. Buchberger et al.[118] described the use of LCEC with a nickel working electrode for the measurement of several monosaccharides and disaccharides. Amperometric detection of reducing carbohydrates, based on coupling with a copper phenanthroline complex as a mediator, was described by Watanabe and Inoue.[119] The method was applied to urine and serum samples; characteristic chromatograms for urine and serum are shown in Figure 4-12. A detection limit of 0.2 ng (1 pmol) of glucose was obtained. In a subsequent work the method was applied for measurements of the amino sugar muramic acid in serum.[120]

Johnson and coworkers[15] described the utility of triple-pulse amperometry at platinum electrodes for the detection of simple alcohols in chromatographic and flow injection systems. Ethanol can be measured in human blood on the basis of its enzymatic reaction with nicotinamide adenine dinucleotide, and monitoring the reduced form of this cofactor.[121] A chromatogram demonstrating the applicability of this method to a blood sample taken from a driver involved in a one-car accident is shown in Figure 4-13. Baldwin and coworkers[122] described indirect amperometric detection of alcohols, based on addition of hydroquinone to the chromatographic mobile phase. The decrease in the hydroquinone oxidation background current was shown to be proportional to the alcohol concentration.

4-1.3.2 LCEC MEASUREMENT OF PHARMACEUTICAL COMPOUNDS IN BIOLOGICAL SAMPLES

Many pharmaceutical compounds possess electroactive functionalities and are easily measurable by LCEC. The determination of antibiotics, alkaloids, anesthetics, and other pharmaceuticals in biological samples has been reported, as the following describes.

Van der Lee and coworkers[123] reported a liquid chromatographic method for determining chloramphenicol and its nitro degradation products. Reductive amperometric detection at a hanging mercury drop electrode, kept at −0.5 V,

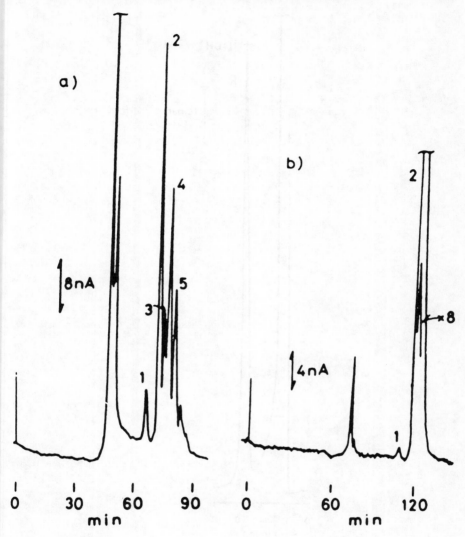

FIGURE 4-12
Chromatograms for (a) urine and (b) serum. Species identity: (a)
lactose (1), glucose (2), galactose (3), fructose (4), arabinose (5);
(b) maltose and lactose (1), glucose(2). (Reproduced with
permission.[119])

yielded a detection limit of 1 ng of chloramphenicol. Application to extracts of
bovine blood was reported. An LCEC procedure for determining theophylline and
its derivatives was described by Greenberg and Mayer.[124] A wax-impregnated
graphite electrode, held at +1.24 V, was employed. The amperometric detection
yielded greater sensitivity and selectivity than 254-nm ultraviolet detection.

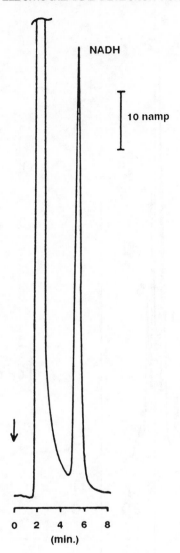

FIGURE 4-13
Chromatogram of a coroner blood sample that was taken from a driver involved in a one-car accident. (Reproduced with permission.[121])

Oxidation of lidocaine and its principal metabolites at a glassy carbon detector was used to detect these compounds in blood serum following chromatographic separation.[125] Detection limits of 2–5 ng were obtained, and total analysis time was 10–15 min.

Reductive differential-pulse amperometric detection was used for measuring chlordiazepoxide and its three major metabolites in plasma.[126] The compounds were selectively extracted into diethyl ether prior to injection. Characteristic chromatograms are illustrated in Figure 4-14. Lund and coworkers[127] evaluated the utility of reductive LCEC for the determination of diazepam, nitrazepam, and chlordiazepoxide. AC polarographic detection was used to monitor nitrazepam in chromatographic effluents.[128] Reductive LCEC, with a static mercury drop electrode, was used to determine the binding of diazepam to human serum albumin.[129]

LCEC procedures for measuring acetaminophen and its metabolites in biological fluids have been reported.[1, 130, 131] These include measurements of acetaminophen in serum at the 50-ng/ml level using liquid chromatography with a pellicular polyamide packing,[1] as well as various procedures for measuring acetaminophen metabolites in urine.[1, 130, 131] The amperometric detection yielded a substantial lowering of detection limits when compared with simultaneous UV detection.[130] Because urinary metabolites of acetaminophen differ widely in oxidation potentials, dual-electrode operation—in the parallel mode—can provide improvement in the qualitative and quantitative information about eluting peaks (Figure 4-8).[38] LCEC with a strong anion-exchange resin, an acetate buffer as an aqueous mobile phase, and a thin-layer amperometric detector was applied to measurement of salicyluric acid in urine following aspirin dose.[132]

The electroactivity of phenothiazine compounds makes them suitable for measurement by LCEC. Wallace et al.[133] reported an LCEC procedure for the determination of promethazine and other phenothiazine compounds in serum. The method requires only a single extraction of serum with 0.2% methanol in hexane, and allows measurement of subtherapeutic concentrations (a detection limit of 0.2 ng/ml). Other effective LCEC methods for the determination of phenothiazine compounds in plasma have been reported.[134–137] The LCEC method provided various advantages over gas chromatography–mass spectrometry for pharmacokinetics studies of promethazine.[134] Trifluoperazine can be measured in biological fluids using dual-electrode detection.[136]

Other drugs used in the treatment of psychotic disorders can be measured by LCEC. Kissinger et al.[138] described measurement of L-dopa in serum, using oxidative LCEC (at +0.77 V). A detection limit of 10 ng/ml was reported. The antidepressants imipramine and desipramine, and their 2-hydroxy metabolites, can be detected in plasma, based on their oxidation at a glassy carbon electrode.[139, 140] The procedure requires a simple solvent extraction, based on the acid-base characteristics of these compounds.

Various LCEC procedures for measuring morphine in body fluids have been reported.[141–143] White[141] used LCEC for measuring morphine and its major metabolite, morphine-3-glucuronide, in blood. The use of oxidative LCEC for the determination of morphine antagonists in urine and plasma was described by Derendorf et al.[142] Optimal oxidation potentials were 0.65 V for morphine and 0.75 V for naloxone and nalrexone. Morphine and naloxone were measured in rat brain using LCEC after a rapid extraction procedure.[143] Oxidative LCEC was used to measure codeine and other narcotic analgesics in urine and serum samples.[144]

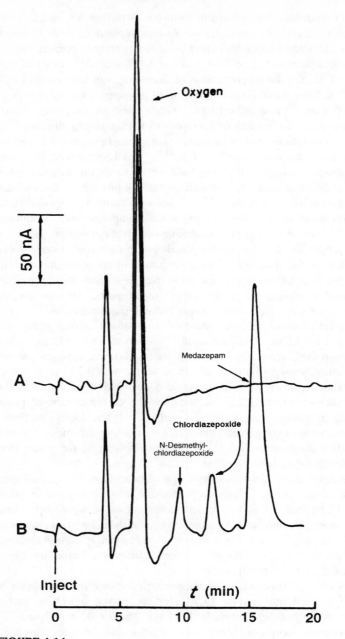

FIGURE 4-14
Liquid chromatography with differential-pulse polarographic
detection. (A) Control plasma; (B) control plasma containing 100
ng/m*l* chlordiazepoxide and *N*-desmethylchlordiazepoxide with 1.5
μg/m*l* medazepam. (Reproduced with permission.[126])

Schwartz and David[145] described an LCEC method for the determination of opium alkaloids, heroin, and cocaine, based on the oxidation of the aliphatic tertiary nitrogen atom common to these compounds. Detection limits of 0.3 ng morphine and 1 ng heroin were obtained.

Various groups reported the usefulness of LCEC for the determination of anticancer platinum complexes in body fluids.[146–149] Both reductive and oxidative LCEC can be used to monitor these agents in cancer patient samples. The reductive mode offered significant advantages over UV and atomic absorption detections.[146] The detector selectivity toward *cis*-dichlorodiammineplatinum(II) allowed its direct measurement in untreated urine at levels below 100 ng/ml. Figure 4-15 compares electrochemical and UV chromatograms of a urine sample collected after administering the drug. Oxidative LCEC measurements of this drug in serum yielded a detection limit of 1 μg/ml.[148] Reductive differential-pulse polarographic detection of the antitumor platinum complex carboplatin following chromatographic separation was described by Elferink and coworkers.[149] Detection limits of 0.1 and 1 μM were obtained for carboplatin in plasma and urine, respectively.

LCEC procedures for monitoring the anthracycline antibiotic doxorubicin (adriamycin) were reported.[150, 151] Complementary information (of the identity of the peaks in the serum extracts) was obtained by using fluorescence detection in conjunction with the amperometric monitoring.[151] Extremely low detection limits of 10 ng/ml were obtained following modest treatment of the blood and urine samples.[150] Palmisano et al.[152] described a sensitive procedure for the determination of the antineoplastic agent methotrexate in human serum and urine; the procedure was based on liquid chromatography with amperometric detection at a glassy carbon electrode held at +0.95 V. The antitumor agent mitomycin C can be measured in plasma, serum, and urine by reversed-phase liquid chromatography with electrochemical detection.[153] El-Yazigi and Martin[154] described a rapid and sensitive LCEC method for routine quantitation of etoposide in the plasma of cancer patients undergoing chemotherapy. A simple one-step extraction procedure was employed. Modified carbon paste electrodes allowed electrocatalytic amperometric detection of thiopurines in blood plasma after liquid chromatography.[155]

Because of the high potency of various cardiovascular drugs, they are usually administered in small doses, resulting in low concentrations in body fluids. Thus, extremely sensitive and selective analytical procedures are required for monitoring these drugs. LCEC has been applied successfully for the determination of various cardioactive drugs. For example, low concentrations of several β-adrenergic blocking drugs—used in the treatment of hypertension and angina—have been measured amperometrically. These include the determination of timolol,[156] pindolol,[157] salbutamol,[158] labetalol,[159] and clenbuterol[160] in human plasma or urine. Bratin[161] described the utility of reductive LCEC for measuring the calcium blocking agent nifedipine. The method is suited for measuring the drug, its metabolites, and photodecomposition products. The structurally similar calcium blocker nicardipine can also be monitored in flowing streams using amperometric detection.[162] Tensammetric detection at a mercury electrode can be used to detect various cardiac glycosides, including digoxin, following reversed-phase chro-

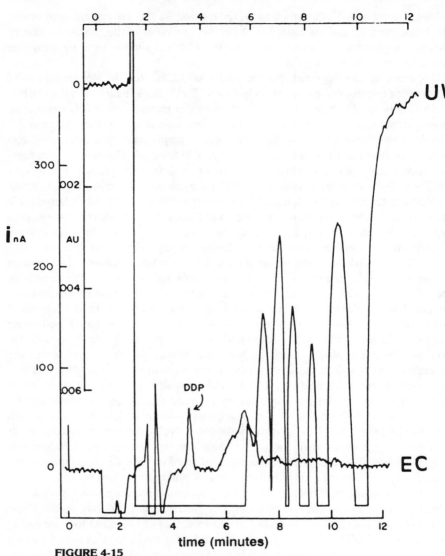

FIGURE 4-15
Electrochemical and ultraviolet chromatograms of patient urine
taken 4 min after administration of cis-dichloroammineplatinum (II).
(Reproduced with permission.[146])

matographic separation.[95] LCEC procedures for monitoring other cardioactive
drugs, such as yohimbine, ibopamine, and fenoldopam, in body fluids have been
reported.[163-165] LCEC at glassy carbon and mercury film thin-layer detectors can
be used to determine the antihypertensive agents reserpine[166] and captopril,[167]
respectively, in body fluids. The fate of captopril in patients receiving the drug was
demonstrated for several hours after administration. The determination of hydrala-

zine and its primary metabolites in urine was reported by Ravichandran and Baldwin.[168] Enhanced oxidation was observed at an electrochemically pretreated glassy carbon detector, and quantitation was accomplished without any sample treatment except particulate filtration. LCEC with a glassy carbon detector held at +1.05 V can be used to detect nanogram levels of warfarin.[169]

New advances in LCEC technology, aimed at extending the scope of LCEC toward additional analytes of clinical interest, are expected in the near future. Among the expected developments are the incorporation of new chemically modified electrodes, new detection schemes, and increased utility of precolumn and postcolumn derivatization reactions.

4-2 ELECTROCHEMICAL DETECTION FOR AUTOMATED FLOW SYSTEMS—CLINICAL ASPECTS

Automation within the clinical laboratory has been a major source of improvements in speed, reproducibility, cost, simplicity, and versatility. Two types of automated flow methods are commonly used: segmented flow methods, in which the analytical stream is divided into discrete segments by air bubbles, and (the relatively new) flow injection analysis (FIA), which utilizes an unsegmented flowing stream into which reproducible sample volumes are injected. These methodologies have contributed toward a revolution in health-care practice. Sampling rates over 120/h are common with both methods. Several reviews of these methods are available.[170–173]

Applications of electrochemical detectors in automated flow systems have gained popularity in recent years. Such detectors are being used to an increasing extent for many automated clinical assays. Because automated flow systems usually lack the separation power of chromatographic systems, specific or multiparametric detection modes are commonly used. These include various species-selective sensors (such as ion-selective electrodes, enzyme electrodes, immunoelectrodes), as well as fast potential scanning techniques. Various automated systems provide an on-line sample pretreatment (e.g., extraction, dialysis) aimed to isolate the analyte; the inherently sensitive (and nonspecific) amperometric detection can be used in conjunction with these systems. The following sections review the clinical applications of automated flow systems with electrochemical detection.

4-2.1 Potentiometric Detection for Automated Flow Systems

Potentiometric ion-selective electrodes have become widely used as detectors in automated flow systems for the determination of clinically important ions, substrates, or enzymes in complex biological systems. Many clinical analyzers used in large hospitals are based on flow systems with potentiometric detection. Analytical results obtained with ion-selective electrodes in flowing streams can be greatly affected by the design of the flow-through electrode. Various designs were proposed, as was described on page 120. Despite the relatively slow response of ion-selective electrodes, high sampling rates have been obtained. Air-segmented and flow injection systems, with potentiometric detection, reaching 360 and 720 samples per hour, respectively, have been reported.[174, 175] Theoretical and practical aspects of this area have been reviewed.[176, 177] The dispersion of the sample plug within the flow injection system results in inferior dynamic measuring ranges and higher detection limits compared with analogous batch measurements.

One of the earliest flow-through assemblies was developed by Technicon for sodium and potassium measurements in the SMAC system.[52] A flow-through sodium-selective glass capillary or valinomycin-based potassium sensor is used with a downstream saturated calomel reference electrode. A comparative study of four commercial ion-selective analyzers (SMAC—Technicon, Astra—Beckman, IL 343—Instrumentation Laboratories, AMT 721—Applied Medical Technology) showed that all four instruments can be used for the determination of sodium and potassium in serum.[178] Meyerhoff and Kovach[179] described a flow injection system with a tubular potassium-selective electrode. The system allows the determination of potassium in serum at a rate of 100 samples per hour. Figure 4-16 shows a trace of a potentiometric recording generated by this system. Virtanen[180] described the use of multiple flow-through ion-selective electrodes in sequence for the simultaneous determination of sodium, potassium, calcium, and chloride ions in serum. Flow-through cells with ion-sensitive field effect transistors (ISFETs) have been proposed for simultaneous measurements of potassium, calcium, and hydrogen

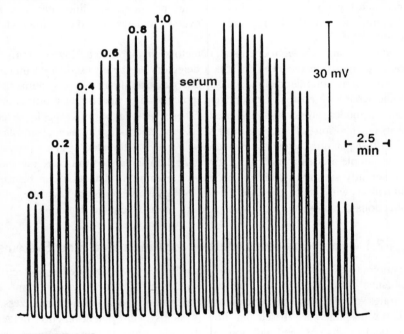

FIGURE 4-16
Flow injection potentiometric determination of potassium in serum.
(Reproduced with permission.[179])

ions.[181, 182] The use of ISFETs permits substantial miniaturization of the system and reduced reagent consumption. Ruzicka et al.[183] described a novel flow injection system for rapid (125 samples per hour) determination of sodium and potassium in blood. Comparison with flame photometric detection yielded similar results. Xie and Christian[184] incorporated coated-wire lithium ion-selective electrodes in a flow injection dialysis system. Typical electrode response signals, using undiluted serum standards and samples (200 μl), are shown in Figure 4-17.

The use of ion-selective electrodes for continuous-flow gas measurements has received considerable attention. Meyerhoff et al.[185] described an automated flow system for potentiometric sensing of ammonia and carbon dioxide. Rapid response times were achieved by using a pH-responsive polymer membrane electrode (instead of a conventional pH glass membrane). Figure 4-18 shows a schematic of the manifold arrangement (including a simple flow-through gas dialysis device) used for automatic determination of ammonia in blood samples. Stewart et al.[186] used an air-gap electrode for potentiometric detection of ammonia. In this approach, the sample is injected into a stream of sodium hydroxide. The ammonia gas subsequently evolved diffuses across the air gap to a pH-sensitive glass electrode to assay about 100 samples per hour.

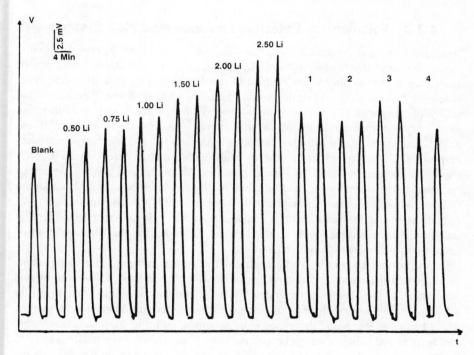

FIGURE 4-17
Response of a lithium-selective electrode for serum standards and serum samples; concentration units are in mM. (Reproduced with permission.[184])

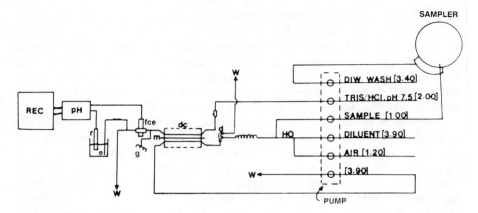

FIGURE 4-18
Automated flow system for the determination of ammonia-N in blood samples. (REC) recorder, (pH) pH-mV meter, (r) reference electrode, (e) electrolyte solution, (fce) ion-selective electrode, (m) gas-permeable membrane, (dc) dialysis membrane. (Reproduced with permission.[185])

4-2.2   Voltammetric Detection for Automated Flow Systems

Amperometric (fixed-potential) detection for automated flow systems is suitable mainly for rapid assays of simple samples. Complex biological samples may benefit from such detection only when coupled to specific enzymatic or immunochemical reactions; this coupling will be described later in the chapter. Some examples of the utility of amperometric detection in automated flow systems include high-speed flow injection measurements of ascorbic acid,[187] isoniazid,[188] and penicilloic acid[189] using reticulated vitreous carbon, glassy carbon, and mercury electrodes, respectively. Automated flow systems with amperometric detection have been useful for the determination of various proteins. Alexander and Shah[190] described an automated method for the determination of albumin in blood serum. The method utilized the shift in the polarographic half-wave potential of potassium titanium(IV) oxalate after reaction with albumin. Yuan and Huber[191] described a flow injection amperometric technique, with a nickel oxide electrode, for the determination of proteins and for denaturation studies.

Fast potential scanning voltammetric detection has received increasing attention for multispecies detection in automated flow systems. Differential-pulse voltammetry has been incorporated with segmented-air and flow injection systems.[192, 193] Several electroactive drugs were measured simultaneously, at rates of 15–20 samples per hour. Faster scanning schemes, and thus high sampling rates, can be obtained using rapid cyclic voltammetry[194] or square-wave polarography[195] in flow systems. Simultaneous detection of several trace metals can be accomplished by coupling stripping voltammetry with automated flow systems. The design and characterization of segmented-air and flow injection systems for automated stripping analysis have been described.[196–199] A recent review summarizes the development of on-line stripping systems.[200] Besides the speed and automation capabilities, on-line manipulations, particularly the medium-exchange approach, can be used to avoid potential interferences (e.g., oxygen) and to improve the resolution.[197, 198] The medium-exchange scheme can also be coupled with adsorptive stripping analysis in flow injection systems to yield selective determination of drugs, such as chlorpromazine or adriamycin, in untreated urine samples.[201, 202]

4-2.3 Electrochemical Detection, Coupled With Specific Biochemical Reactions, for Flow Analysis

The coupling of enzymatic reactions with automated flow systems offers selective determination, with high sampling rates and low cost. One of the first (and most important) examples of such coupling is the use of glucose oxidase for automated glucose assays. The details of the amperometric detection of the liberated hydrogen peroxide are given in Chapter 3. Llenado and Rechnitz[51] described an air-segmented automated flow system for the enzymatic determination of glucose. The system functioned well in conjunction with serum samples containing glucose in the physiological concentration range at sampling rates up to 70 determinations per hour. Watson et al.[203] described a dedicated instrument for the determination of glucose in serum or plasma. The flow system included a column of immobilized glucose oxidase and an amperometric sensor to measure the hydrogen peroxide. With $2\text{-}\mu l$ samples and no sample preparation, the sampling rate was 60/h. Commercial glucose flow analyzers, based on similar principles, are available from several companies. Glucose assays in flowing streams can be performed using a chemically modified amperometric electrode, which offers insensitivity to the oxygen level[204]; with $50\mu l$ samples, detection limits of $0.25\ \mu M$ of glucose can be achieved.

In addition to glucose assays, automated flow systems with enzymatic reactions and electrochemical detection have been applied successfully to measurements of other clinically important substrates. Gnanasekaran and Mottola[205] described the flow injection determination of penicillins using immobilized penicillinase in a single bead string reactor. Selective measurements were obtained at the $50\text{--}500\mu M$ range, at a frequency of 150 injections per hour. Olson's group[206, 207] described the utility of enzymatic differential amperometry at tubular carbon electrodes in flowing streams for the determination of cholesterol and alcohol in body fluids. Llenado and Rechnitz[208] designed an automatic flow system with an ammonia-sensing membrane probe for the enzymatic determination of urea. Mascini and Palleschi[209] used immobilized creatininase in conjunction with a flow-through ammonia electrode for the automated determination of creatinine in body fluids. The same group described the simultaneous flow analysis of urea and glucose in serum samples, using dual enzyme electrodes.[210] Flow injection analysis of lactate can be performed using a modified carbon electrode covered with a lactate oxidase membrane.[211] The linear range is $0.01\text{--}3$ mM, and the assay rate is 200 samples per hour. Blaedel and Engstrom[212] developed a flow system with a reagentless enzyme electrode for the flow analysis of plasma ethanol. The system offers high sensitivity, rapid response time, and good accuracy and reproducibility.

Immunochemical separation offers another approach to improve the specificity of flow analyzers with electrochemical detection. Various flow-through immunoassay procedures, based on amperometric or potentiometric detection, have been reported. Such flow-through immunoassay procedures enable carefully controlled conditions, and offer rapid automation coupled with highly sensitive detection. The relatively simple instrumentation requirements are advantageous compared with other types of immunoassay. One effective approach involves enzyme immunoassay with amperometric detection. Enzyme immunoassay is based on labeling the antigen with an enzyme that catalyzes the production of an electroactive product. Because every mole of enzyme can produce at least 10^3– 10^4 moles of product, an amplification is obtained. Detection limits are often restricted by the antibody-antigen binding constant rather than the electrochemical technique. Eggers et al.[213] demonstrated the utility of enzyme immunoassay of phenytoin, using glucose-6-phosphate dehydrogenase as the label and amperometric detection of the NADH produced. Using serum samples, good agreement was obtained with an established clinical procedure for phenytoin. De Alwis and Wilson[214] described a rapid enzyme-linked immunosorbent assay for immunoglobulin G. In this high-performance immunoaffinity chromatographic (HPIC) scheme the antibody was immobilized on an immunoaffinity column (Figure 4-19). The antigen labeled with enzyme and the antigen (in the injected sample) competed for the binding with the antibody. This was followed by injection of the enzyme substrate. The method

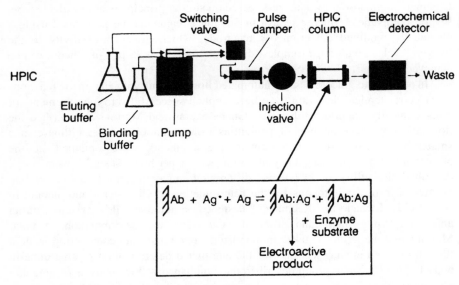

FIGURE 4-19
Experimental setup for high-performance immunoaffinity chromatography with electrochemical detection. (Reproduced with permission.[214])

involved a minimum incubation time (0.1 min) while offering subpicomole sensitivity. The immunosorbent reactor was shown to be stable for 500 repetitive injections over a 3-month period. Glucose oxidase was used as the enzyme label and hydrogen peroxide as the sensed material. Heineman's group[215] utilized the inherent sensitivity of amperometric detection for phenol as a basis for effective heterogeneous enzyme immunoassay for digoxin. A detection limit of 50 pg/ml of digoxin was achieved in plasma samples. Retardation of phenol on a short octyldecylsilane column separated its peak from interfering peaks. Comparison with radioimmunoassay yielded a good correlation ($r = 0.95$). Typical response peaks for a series of digoxin standards in plasma solutions are shown in Figure 4-20. Similarly,

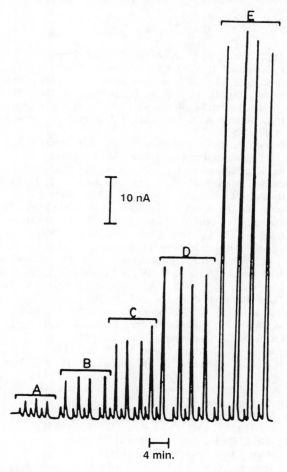

FIGURE 4-20
Heterogeneous enzyme immunoassay with LCEC detection for a series of digoxin standards in plasma solutions. Concentration of digoxin in plasma samples was (A) 5.0, (B) 2.0, (C) 1.0, (D) 0.5, and (E) 0.0 ng/ml. (Reproduced with permission.[215])

enzymatic production and amperometric detection of phenol can be used in the immunoassay of α_1-acid glycoprotein.[216] The same group reported recently a much faster enzyme immunoassay for digoxin based on liquid chromatographic column switching with amperometric detection of the NADH produced.[217] Guilbault's group[218, 219] described flow systems for rapid and convenient measurements of the isoenzymes lactate dehydrogenase and creatine kinase MB in serum. The isoenzymes were separated immunochemically, with potassium ferricyanide used as a mediator. A review by Sittampalam and Wilson[220] covers the principles and applications of flow-through enzyme immunoassays with electrochemical detection.

REFERENCES

1. Riggin, R. M., Schmidt, A. L., Kissinger, P. T., *J. Pharm. Sci.*, *64*, 680 (1975).
2. Koch, D. D., Kissinger, P. T., *J. Chromatogr.*, *164*, 441 (1979).
3. Lemar, M., Porthault, M., *J. Chromatogr.*, *130*, 372 (1977).
4. Gunasingham, H., Tay, B. T., Ang, K. P., *Anal. Chem.*, *56*, 978 (1984).
5. Bond, A.M., Fleischmann, M., Robinson, J., *J. Electroanal. Chem.*, *168*, 299 (1984).
6. Anderson, J. L., Whiten, K. K., Brewster, J. D., Ou, T. Y., Nonidez, W. K., *Anal. Chem.*, *57*, 1366 (1985).
7. Wang, J., Dewald, H. D., *J. Chromatogr.*, *285*, 281 (1984).
8. Curran, D. J., Tougas, T. P., *Anal. Chem.*, *56*, 672 (1984).
9. Ravichandran, K., Baldwin, R. P., *Anal. Chem.*, *55*, 1782 (1983).
10. Ravichandran, K., Baldwin, R. P., *J. Liq. Chromatogr.*, *7*, 2031 (1984).
11. Wang, J., Hutchins, L. D., *Anal. Chem.*, *57*, 1536 (1985).
12. Hutchins-Kumar, L. D., Wang, J., Tuzhi, P., *Anal. Chem.*, *58*, 1019 (1986).
13. Wang, J., Golden, T., Tuzhi, P., *Anal. Chem.*, *59*, 740 (1987).
14. Hughes, S., Johnson, D. C., *Anal. Chim. Acta*, *149*, 1 (1983).
15. Hughes, S., Meschi, P. L., Johnson, D. C., *Anal. Chim. Acta*, *132*, 1 (1981).
16. Buchbenger, W., Winsauer, K., Breitwiesser, C., *Z. Anal. Chem.*, *315*, 518 (1983).
17. Kok, W., Brinkman, U. A., Frei, R. W., *J. Chromatogr.*, *256*, 17 (1983).
18. Weber, S. G., Purdy, W. C., *Ind. Eng. Chem. Prod. Res. Dev.*, *20*, 593 (1981).
19. Weber, S. G., *J. Electroanal. Chem.*, *145*, 1 (1983).
20. Stulik, K., Pacakova, V., *J. Electroanal. Chem.*, *129*, 1 (1981).
21. "Application Note C-3," EG&G PAR Inc., Princeton, N.J.
22. Slais, K., Krejci, M., *J. Chromatogr.*, *235*, 21 (1982).
23. Jin, Z., Rappoport, S. M., *Anal. Chem.*, *55*, 1778 (1983).
24. Hirata, Y., Lin, P. T., Novotny, M., Wightman, R. M., *J. Chromatogr.*, *181*, 287 (1980).
25. Knecht, L. A., Guthrie, E. J., Jorgenson, J. W., *Anal. Chem.*, *56*, 479 (1984).
26. Rucki, R. J., *Talanta*, *27*, 147 (1980).
27. MacCrehan, W. A., *Anal. Chem.*, *53*, 74 (1981).
28. Johnson, D. C., Hsi, T., *Anal. Chim. Acta*, *175*, 23 (1985).
29. Edwards, P., Haak, K. K., *Am. Lab.*, *15* (4), 78 (1983).
30. Last, T. A., *Anal. Chim. Acta*, *155*, 287 (1983).
31. Wang, J., Ouziel, E., Yarnitzky, C., Ariel, M., *Anal. Chim. Acta*, *102*, 99 (1978).
32. Samuelsson, R., Osteryoung, J., *Anal. Chim. Acta*, *123*, 97 (1981).
33. Reardon, P. A., Obrien, G. E., Sturrock, P. E., *Anal. Chim. Acta*, *162*, 175 (1984).
34. Trojanek, A., De Jong, H. G., *Anal. Chim. Acta*, *141*, 115 (1982).
35. Caudill, W., Ewing, A. G., Jones, S., Wightman, R. M., *Anal. Chem.*, *55*, 1877 (1983).

36. Last, T. A., *Anal. Chem., 55,* 1509 (1983).
37. Trubey, R. K., Nieman, T. A., *Anal. Chem., 58,* 2549 (1986).
38. Roston, D. A., Shoup, R. E., Kissinger, P. T., *Anal. Chem.,* 54, 1417A (1982).
39. Andrews, R. W., Schubert, C., Morrison, J., Zink, E. W., Matson, W. R., *Am. Lab., 14* (10), 140 (1982).
40. Allison, L. A., Shoup, R. E., *Anal. Chem., 55,* 8 (1983).
41. Lunte, C. E., Kissinger, P. T., *Anal. Chem., 55,* 1458 (1983).
42. Johnson, D. C., Weber, S. G., Bond, A. M., Wightman, R. M., Shoup, R. E., Krull, J. S., *Anal. Chim. Acta, 180,* 187 (1986).
43. Kissinger, P. T., *Anal. Chem., 49,* 447A (1977).
44. Svoboda, V., Marsal, J., *J. Chromatogr., 148,* 111 (1978).
45. Tesarik, K., Kalab, P., *J. Chromatogr., 78,* 357 (1973).
46. Small, H., Stevens, T. S., Bauman, W. C., *Anal. Chem., 47,* 1801 (1975).
47. Anderson, C., *Clin. Chem., 22,* 1424 (1976).
48. Haderka, S., *J. Chromatogr., 54,* 357 (1971).
49. Alder, J. F., Drew, P. K. P., Fielden, P. R., *Anal. Chem., 55,* 256 (1983).
50. van der Linden, V., Oostervink, R., *Anal. Chim. Acta, 101,* 419 (1978).
51. Llenado, R. A., Rechnitz, G. A., *Anal. Chem., 45,* 2165 (1973).
52. Rao, K. J., Pelavin, M. H., Morgenstern, S., *Advances in Automated Analysis,* 1972 Technicon International Congress, vol. 1, pp. 33–36.
53. Loscombe, C. R., Cox, G. B., Dabziel, J. A. W., *J. Chromatogr., 166,* 403 (1978).
54. Kissinger, P. T., *J. Chem. Educ., 60,* 308 (1983).
55. Church, W. H., Justice, J. B., *Anal. Chem., 59,* 712 (1987).
56. Riggin, R. M., Kissinger, P. T., *Anal. Chem., 49,* 2109 (1977).
57. Oke, A., Keller, R., Mefford, I., Adams, R. N., *Science, 200,* 1411 (1978).
58. Smith, J. E., Lane, J. D., *Pharm. Biochem. Behavior, 16,* 641 (1982).
59. Mefford, I. N., Foutz, A., Noyce, N., Jurik, S. M., Handen, C., Dement, W. C., Barchas, J. D., *Brain Res., 236,* 339 (1982).
60. Yamamoto, T., Shimizu, H., Kato, T., Nagatsu, T., *Anal. Biochem., 142,* 395 (1984).
61. Wagner, J., Vitali, P., Palfreyman, M. G., Zraika, M., Hout, S., *J Neurochem., 38,* 1241 (1982).
62. Kim, C., Campanelli, C., Khanna, J. M., *J. Chromatogr., 282,* 151 (1983).
63. Harris, P. Q., Bacopoulos, N. G., Brown, S. J., *J. Chromatogr., 309,* 379 (1984).
64. Matson, W. R., Langlais, P., Volicer, L., Gamache, P. H., Bird, E., Mark, K. A., *Clin. Chem., 30,* 1477 (1984).
65. Hefti, F., *Life Sciences, 25,* 775 (1979).
66. Zoutendam, P. H., Bruntlett, C. S., Kissinger, P. T., *Anal. Chem., 48,* 237 (1976).
67. Davis, G. C., Kissinger, P. T., *Anal. Chem., 51,* 1960 (1979).
68. Blank, C. L., Pike, R., *Life Sciences, 18,* 859 (1976).
69. Shoup, R. E., Bruntlett, C. S., Jacobs, W. A., Kissinger, P. T., *Am. Lab., 13* (10), 144 (1981).
70. Kissinger, P. T., Bruntlett, C. S., Shoup, R. E., *Life Sciences, 28,* 455 (1981).
71. Rice, J. R., Kissinger, P. T., *J. Anal. Toxicol., 3,* 64 (1979).
72. Radzik, D. M., Roston, D. A., Kissinger, P. T., *Anal. Biochem., 131,* 458 (1983).
73. Fiala, E. S., Kulakis, C., *J. Chromatogr., 214,* 229 (1981).
74. Rabenstein, D. L., Saetre, R., *Anal. Chem., 49,* 1036 (1977).
75. Rabenstein, D. L., Saetre, R., *Clin. Chem., 24,* 1140 (1978).
76. Stein, A. F., Dills, R. L., Klaassen, C. D., *J. Chromatogr., 381,* 259 (1986).
77. Shaw, I. C., McLean, A. E. M., Boult, C. H., *J. Chromatogr., 275,* 206 (1983).
78. Halbert, M. K., Baldwin, R. P., *J. Chromatogr., 345,* 43 (1985).
79. Lunte, C. E., Kissinger, P. T., *Anal. Chim. Acta, 158,* 33 (1984).
80. Pachla, L. A., Kissinger, P. T., *Clin. Chim. Acta, 59,* 309 (1975).
81. Slaunwhite, W. D., Pachla, L. A., Wenke, D. C., Kissinger, P. T., *Clin. Chem., 21,* 1427 (1975).

82. Iwamoto, T., Yoshiura, M., Iriyama, K., *J. Chromatogr., 278,* 156 (1983).
83. Palmisano, F., Desimoni, E., Zambonin, P. G., *J. Chromatogr., 306,* 205 (1984).
84. Kafil, J. B., Cheng, H. Y., Last, T. A., *Anal. Chem., 58,* 285 (1986).
85. Stulik, K., Pacakova, V., *J. Chromatogr., 273,* 77 (1983).
86. Kok, W. T., Groenendijk, F., Brinkman, V. A. T., Frei, R. W., *J. Chromatogr., 315,* 271 (1985).
87. Mayer, W. J., Greenberg, M. S., *J. Chromatogr. Sci., 17,* 614 (1979).
88. Mayer, W. J., McCarthy, J. P., Greenberg, M. S., *J. Chromatogr. Sci., 17,* 656 (1979).
89. Hepler, B. R., Weber, S. G., Purdy, W., *Anal. Chim. Acta, 113,* 269 (1980).
90. Shihabi, Z. K., Scaro, J., Thomas, B. F., *J. Chromatogr., 224,* 99 (1981).
91. "LCEC Application Note No. 35," Bioanalytical Systems Inc., W. Lafayette, Ind. (1982).
92. Gunasingham, H., Tay, B. T., Ang, K. P., *J. Chromatogr., 341,* 271 (1985).
93. Kenyhercz, T. M., Kissinger, P. T., *J. Anal. Toxicol., 2,* 1 (1978).
94. Smyth, M. R., Frischkorm, C. G. B., *Z. Anal. Chem., 301,* 220 (1980).
95. De Jong, H. G., Voogt, W. H., Bos, P., Frei, R. W., *J. Liq. Chromatogr., 6,* 1745 (1983).
96. Kutner, W., Debowski, J., Kemula, W., *J. Chromatogr., 191,* 47 (1980).
97. Shimada, K., Tanaka, M., Nambara, T., *Anal. Lett., 13,* 1129 (1980).
98. Mason, W. D., Amick, E. N., Heft, W., *Anal. Lett., 13,* 817 (1980).
99. Pachla, L. A., Kissinger, P. T., *Anal. Chem., 48,* 364 (1976).
100. Heiliger, F. C., *Current Separations (BAS), 2*(3), 4 (1980).
101. Dozier, J. C., Salamone, J. D., Neill, D. B., Justice, J. B., *Current Separations (BAS), 4*(1), 1 (1982).
102. Carr, R. S., Neff, J. M., *Anal. Chem., 52,* 2428 (1980).
103. Brunt, K., Bruins, C. H. P., *J. Chromatogr., 172,* 37 (1979).
104. Tsao, C. S., Young, M., *J. Chromatogr., 330,* 408 (1985).
105. Hart, J., Shearer, M. J., McCarthy, P. T., *Analyst, 110,* 1181 (1985).
106. Haroon, J., Schubert, C. A. W., Hauschka, P. V., *J. Chromatogr. Sci., 22,* 89 (1984).
107. Chou, P. P., Jaynes, P. K., Bailey, J. L., *Clin. Chem., 31,* 880 (1985).
108. Vandewoude, M., Claeys, M., De Leeuw, I., *J. Chromatogr., 311,* 176 (1984).
109. Huang, M., Burckart, G. J., Venkatoramanan, R., *J., Chromatogr., 380,* 331 (1986).
110. Allenmark, S., Hjelm, E., Larsson-Cohn, L., *J. Chromatogr., 146,* 485 (1978).
111. Polta, J. A., Johnson, D. C., *J. Liq. Chromatogr., 6,* 1727 (1983).
112. Allison, L. A., Mayer, G. S., Shoup, R. E., *Anal. Chem., 56,* 1089 (1984).
113. Mahachi, T. J., Carlson, R. M., Poe, D. P., *J. Chromatogr., 298,* 279 (1984).
114. Wightham, R. M., Paik, E. C., Borman, S., Dayton, M. A., *Anal. Chem., 50,* 1410 (1978).
115. White, M. W., *J. Chromatogr., 262,* 420 (1983).
116. Meek, J. L., *J. Chromatogr., 266,* 401 (1983).
117. Alexander, P. W., Hadda, P. R., Low, G. K. C., Maitra, C., *J. Chromatogr., 209,* 29 (1981).
118. Buchberger, W., Winsaner, K., Breitwieser, C., *Z. Anal. Chem., 315,* 518 (1983).
119. Watanabe, N., Inoue, M., *Anal. Chem., 55,* 1016 (1983).
120. Watanabe, N., *J. Chromatogr., 316,* 496 (1984).
121. Davis, G. C., Holland, K. L., Kissinger, P. T., *J. Liq. Chromatogr., 2,* 663 (1979).
122. Jiannong, Y., Baldwin, R. P., Ravichandran, K., *Anal. Chem., 58,* 2337 (1986).
123. Van der Lee, J. J., Van der Lee-Rijsbergen, Tjaden, U. R., Van Bennekom, W. P., *Anal. Chim. Acta, 149,* 29 (1983).
124. Greenberg, M. S., Mayer, W. J., *J. Chromatogr., 169,* 321 (1979).
125. Halbert, M. K., Baldwin, R. P., *J. Chromatogr., 306,* 269 (1984).
126. Hackman, M. R., Brooks, M. A., *J. Chromatogr., 222,* 179 (1981).
127. Lund, W., Hannisdal, M., Greibrokk, T., *J. Chromatogr., 173,* 249 (1979).

128. Hanekamp, H. B., Voogt, W. H., Frei, R. W., Bos, P., *Anal. Chem., 53,* 1362 (1981).
129. Thuaud, N., Sebille, B., Livertoux, M. H., Bessiere, J. J., *J. Chromatogr., 282,* 509 (1983).
130. Wilson, J. M., Slattery, J. T., Forte, A. J., Nelson, S. D., *J. Chromatogr., 227,* 453 (1982).
131. Hamilton, M., Kissinger, P. T., *Anal. Biochem., 125,* 969 (1982).
132. Kissinger, P. T., Felice, L. J., King, W. P., Pachla, L. A., Riggin, R. M., Shoup, R. E., *J. Chem. Educ., 54,* 50 (1977).
133. Wallace, J. E., Shimek, E. L., Stavchansky, S., Harries, S. C., *Anal. Chem., 53,* 960 (1981).
134. Leelavathi, D. E., Dressler, D. E., Soffer, E. F., Yachetti, S. D., Knowles, J. A., *J. Chromatogr., 339,* 105 (1985).
135. "LCEC Application Note No. 41," Bioanalytical Systems Inc. W. Lafayette, Ind. 1982.
136. "Coulochem Applications—Trifluoperazine," Environmental Sciences Associates Inc., Bedford, Mass. (1982).
137. Melethil, S., Dutta. A., Chungi, V., Dittert, L., *Anal. Lett., 16,* 701 (1983).
138. Kissinger, P. T., Felice, L. J., Riggin, R. M., Pachla, L. A., Wenke, D. C., *Clin. Chem., 20,* 992 (1974).
139. Suckow, R. F., Cooper, T. B., *J. Pharm. Sci., 70,* 257 (1981).
140. "LCEC Application Note No. 40," Bioanalytical Systems Inc., W. Lafayette, Ind. 1982.
141. White, M. W., *J. Chromatogr., 178,* 229 (1979).
142. Derendorf, H., El-Din, A., El-Koussi, A., Garrett, E. G., *J. Pharm. Sci., 73,* 621 (1984).
143. Raffa, R. B., O'Neill, J. J., Tallarida, R. J., *J. Chromatogr., 238,* 515 (1982).
144. Meinsma, D. A., Kissinger, P. T., *Current Separations (BAS), 6,* 42 (1985).
145. Schwartz, R. S., David, K. O., *Anal. Chem., 57,* 1362 (1985).
146. Bannister, S. J., Sternson, L. A., Repta, A. J., *J. Chromatogr., 273,* 301 (1983).
147. Krull, I. S., Ding, X. D., Braverman, S., Selaka, C., Hochberg, F., Sternson, L. A., *J. Chromatogr. Sci., 21,* 166 (1983).
148. Richmond, W. M., Baldwin, R. P., *Anal. Chim. Acta, 154,* 133 (1983).
149. Elferink, F., van der Vijgh, W. J. F., Pinedo, H. M., *Anal. Chem., 58,* 2293 (1986).
150. Akpofure, C., Riley, C. A., Sinkule, J. A., Evans, W. E., *J. Chromatogr., 232,* 377 (1982).
151. Kotake, A. N., Vogelzang, N. J., Larson, R. A., Choporis, N., *J. Chromatogr., 337,* 194 (1985).
152. Palmisano, F., Cataldi, T. R. I., Zambronin, P. G., *J. Chromatogr., 344,* 249 (1985).
153. Tjaden, U. R., Langenberg, J. P., Ensing, K., Van Bennekon, W. P., De Bruijn, E. A., Van Oosterom, A. T., *J. Chromatogr., 232,* 355 (1982).
154. El-Yazigi, A., Martin, C. R., *Clin. Chem., 33,* 803 (1987).
155. Halbert, M. K., Baldwin, R. P., *Anal. Chim. Acta, 187,* 89 (1986).
156. Gregg, M. R., Jack, D. B., *J. Chromatogr., 305,* 244 (1984).
157. Diquet, B., Nguyen-Huu, J. J., Boutron, H., *J. Chromatogr., 311,* 430 (1984).
158. Tan, Y. K., Soldin, S. J., *J. Chromatogr., 311,* 311 (1984).
159. Wang, J., Bonakdar, M., Deshmukh, B. K., *J. Chromatogr., 344,* 412 (1985).
160. Diquet, B., Doare, L., Simon, P., *J. Chromatogr., 336,* 411 (1984).
161. Bratin, K., LCEC Symposium, Indianapolis, paper no. 22 (May 1982).
162. Wang, J., Deshmukh, B. K., Bonakdar, M., *Anal. Lett., 18,* 1087. (1985).
163. Diquet, B., Doare, L., Gaudell, G., *J. Chromatogr., 311,* 1984 (1984).
164. Gifford, R., Randolph, W. C., Heineman, F. C., Ziemniak, J. A., *J. Chromatogr., 381,* 83 (1986).

165. Boppana, V. K., Heineman, F. C., Lynn, R. K., Randolph, W. C., Ziemniak, J. A., *J. Chromatogr.*, *317*, 463 (1984).
166. Wang, J., Bonakdar, M., *J. Chromatogr.*, *382*, 343 (1986).
167. Parrett, D., Drury, P. L., *J. Liq. Chromatogr.*, *5*, 97 (1982).
168. Ravichandran, K., Baldwin, R. P., *J. Chromatogr.*, *343*, 99 (1985).
169. Wang, J., Bonakdar, M., *J. Chromatogr.*, *415*, 432 (1987).
170. Snyder, L. R., *Anal. Chim. Acta*, *114*, 3 (1980).
171. Ruzicka, J., *Anal. Chem.*, *55*, 1040A (1983).
172. Stewart, K. K., *Anal. Chem.*, *55*, 931A (1983).
173. Ruzicka, J., Hansen, E. H., *Anal. Chim. Acta*, *179*, 1 (1986).
174. Alexander, P. W., Seegopaul, P., *Anal. Chem.*, *52*, 2403 (1980).
175. Astrom, O., *Anal. Chim. Acta*, *105*, 67 (1979).
176. Toth, K., Fucsko, J., Lindner, E., Feher, Z., Pungor, E., *Anal. Chim. Acta*, *179*, 359 (1986).
177. Trojanowicz, M., Matuszewski, W., *Anal. Chim. Acta*, *138*, 71 (1982).
178. Gross, S. E., Khayam-Bashi, H., *Clin. Chem.*, *28*, 1629 (1982).
179. Meyerhoff, M. E., Kovach, P. M., *J. Chem. Educ.*, *60*, 766 (1983).
180. Virtanen, R., *Anal. Chem. Symp. Ser.*, *8*, 375 (1981).
181. Ramsing, A. V., Janata, J., Ruzicka, J., Levy, M., *Anal. Chim. Acta*, *118*, 45 (1980).
182. Sibbald, A., Whalley, P. D., Covington, A. K., *Anal. Chim. Acta*, *159*, 47 (1984).
183. Ruzicka, J., Hansen, E. H., Zagatto, E. A., *Anal. Chim. Acta*, *88*, 1 (1977).
184. Xie, R. Y., Christian, G. D., *Anal. Chem.*, *58*, 1806 (1986).
185. Meyerhoff, M. E., Fraticelli, Y. M., Greenberg, J. A., Rosen, J., Parks, S. J., Opdycke, W. N., *Clin. Chem.*, *28*, 1973 (1982).
186. Stewart, J. W. B., Ruzicka, J., Bergamin, H., Zagatto, E. A., *Anal. Chim. Acta*, *81*, 371 (1976).
187. Strohl, A. N., Curran, D. J., *Anal. Chem.*, *51*, 1045 (1979).
188. Shah, M. H., Stewart, J. T., *Anal. Lett.*, *16* (B12), 913 (1983).
189. Forsman, U., Karlsson, A., *Anal. Chim. Acta,139*, 133 (1982).
190. Alexander, P. W., Shah, M. H., *Anal. Chem.*, *52*, 1896 (1980).
191. Yuan, C. J., Huber, C. O., *Anal. Chem.*, *57*, 180 (1985).
192. Cullen, L. F., Brindle, M. P., Papariello, G. J., *J. Pharm. Sci.*, *62*, 1708 (1983).
193. Wang, J., Dewald, H. D., *Anal. Chim. Acta*, *153*, 325 (1983).
194. Thogersen, N., Janata, J., Ruzicka, J., *Anal. Chem.*, *55*, 1986 (1983).
195. Wang, J., Ouziel, E., Yarnitzky, C., Ariel, M., *Anal. Chim. Acta*, *102*, 99 (1978).
196. Wang, J., Ariel, M., *Anal. Chim. Acta*, *101*, 1 (1978).
197. Wang, J., Dewald, H. D., Greene, B., *Anal. Chim. Acta*, *146*, 25 (1983).
198. Hu, A., Dessy, R. E., Graneli, A., *Anal. Chem.*, *55*, 320 (1983).
199. Frenzel, W., Bratter, P., *Anal. Chim. Acta*, *179*, 389 (1986).
200. Wang, J., *Am. Lab.*, *15* (7), 14 (1983).
201. Wang, J., Freiha, B. A., *Anal. Chem.*, *55*, 1285 (1983).
202. Chaney, E. N., Baldwin, R. P., *Anal. Chim. Acta*, *176*, 105 (1985).
203. Watson, B., Stifel, D. N., Semersky, F. E., *Anal. Chim. Acta*, *106*, 233 (1979).
204. Appelquist, R., Marko-Varga, G., Gorton, L., Torstensson, A., Johansson, G., *Anal. Chim. Acta*, *169*, 237 (1985).
205. Gnanasekaran, R., Mottola, H. A., *Anal. Chem.*, *57*, 1005 (1985).
206. Hahn, Y., Olson, C. L., *Anal. Chem.*, *51*, 444 (1979).
207. Smith, M. D., Olson, C. L., *Anal. Chem.*, *47*, 1074 (1975).
208. Llenado, R. A., Rechnitz, G. A., *Anal. Chem.*, *46*, 1109 (1974).
209. Mascini, M., Palleschi, G., *Anal. Chim. Acta*, *136*, 69 (1982).
210. Mascini, M., Palleschi, G., *Anal. Chim. Acta*, *145*, 213 (1983).
211. Scheller, F., Schubert, F., Olsson, B., Gorton, L., Johansson, G., *Anal. Lett.*, *19*, 1691 (1986).
212. Blaedel, W. J., Engstrom, R. C., *Anal. Chem.*, *52*, 1691 (1980).

213. Eggers, H. M., Halsall, H. B., Heineman, W. R., *Clin. Chem., 28,* 1848 (1982).
214. de Alwis, W. U., Wilson, G. S., *Anal. Chem., 57,* 2754 (1985).
215. Wehmeyer, K. R., Halsall, H. B., Heineman, W. R., Volle, C. P., Wen Chen, I., *Anal. Chem., 58,* 135 (1986).
216. Doyle, M. J., Halsall, H. B., Heinemann, W. R., *Anal. Chem., 56,* 2355 (1984).
217. Wright, D. S., Halsall, H. B., Heinemann, W. R., *Anal. Chem., 58,* 2995 (1986).
218. Toyoda, T., Kuan, S. S., Guilbault, G. G., *Anal. Lett., 18* (B3), 345 (1985).
219. Toyoda, T., Kuan, S. S., Guilbault, G. G., *Anal. Chem., 57,* 2346 (1985).
220. Sittampalam, G. S., Wilson, G. S., *Trends Anal. Chem., 3*(4), 96 (1984).

CHAPTER 5

In-Vivo Electrochemistry

5-1 INTRODUCTION

There has been increasing interest in recent years in continuous in-vivo monitoring of important clinical variables. Such real-time measurements are highly desirable in intensive-care units and surgery, as well as for long-term bioavailability and metabolic studies. At present, the clinician is usually restricted to analyses in a central laboratory, with all the associated delays. Important information regarding rapid changes in the levels of blood metabolites (e.g., in acutely ill patients) may be missed using such discrete in-vitro measurements. Continuous monitoring by implantable sensors enables a closer surveillance of patients, via a rapid return of the bioanalytical information. This enables the physician to make necessary rapid decisions, thus improving the overall clinical treatment of the patient.

Close surveillance of patients may be accomplished also by withdrawing the blood to a bedside electrochemical analyzer that displays the analytical data with minimum delay. Future technology will result in biofeedback systems (such as the artificial pancreas), with real-time information leading to direct drug release with an associate drug dispenser. Overall, in-vivo measurements may be used to improve both diagnosis and treatment.

The requirements for an ideal in-vivo probe are (1) high sensitivity and selectivity, (2) long-term stability, (3) fast response time, (4) independence of variations (e.g., pH, oxygen tension) in the biological sample, (5) biocompatibility and sterilizability, and (6) small size. The difficulties in fulfilling some of these requirements currently restrict the widespread clinical use of various invasive in-vivo probes.

In spite of the obvious problems associated with in-vivo monitoring, electrodes remain the most useful tool for this purpose. This results from the fact that electrode processes are surface- rather than volume-dependent, and the chemical information is directly converted into an electrical signal. In-vivo electrochemical detection of various cations and gases, namely K^+, Na^+, Ca^{+2}, H^+, O_2, and CO_2, as well as catecholamines, has become feasible. At present, major attention and efforts are being directed toward the development of microsensors for in-vivo monitoring of organic compounds (drugs, metabolites, hormones). Another major challenge is to minimize the interaction between the implant and the biological environment that

149

gradually degrades the analytical performance. The extent of this interaction depends on the nature of the sensor and the implantation site. The objectives and challenges of in-vivo electrochemistry have been reviewed by Pinkerton and Lawson.[1]

There are basically two main branches of in-vivo electrochemistry: potentiometry and voltammetry. The design, operation, and applications of potentiometric and voltammetric microsensors for in-vivo analysis are reviewed in this chapter.

5-2 IN-VIVO VOLTAMMETRY

5-2.1 In-Vivo Voltammetric Electrodes

For in-vivo voltammetric measurements, suitable working electrodes are usually made of carbon or platinum. Reference (miniature Ag-AgCl) and auxiliary (stainless-steel or platinum wires) electrodes are placed in convenient locations. Platinum microelectrodes are usually employed for monitoring oxygen or glucose. Carbon (paste or fiber) microvoltammetric electrodes are commonly used in brain tissues. The carbon paste microelectrode is prepared by packing the paste into the end of a pulled glass capillary or Teflon tubing. For chronic measurements, where a rigid surface is essential, an epoxy resin is added to the carbon paste.[2] The active carbon surface is the tip, 50–200 μm in diameter. Electrical contact to the inside end of the paste is made with a suitable wire. Such electrodes are usually tested and calibrated before implantation. Graphite-epoxy microelectrodes of similar configuration have also been described.[3] Smaller electrodes, made of carbon fibers (5–10-μm diameter) sealed in glass capillaries, have been introduced recently.[4-6] Figure 5-1 shows a schematic of a two-barrel working reference electrode assembly using a 6-μm-diameter carbon fiber. The reduction in electrode size results in a decrease in the number of molecules electrolyzed and reduced tissue damage. A needle-tip carbon fiber electrode (0.5–2-μm tip diameter) has been designed for voltammetric detection of intracellular electroactive compounds.[8] A unique property of electrodes, whose dimensions are on the micrometer scale (smaller than the diffusion-layer thickness), is the sigmoidal obtained in cyclic voltammetry. Such response is a result of enhanced mass transport due to nonlinear diffusion. Modification of microelectrode surfaces has been useful for improving their performance (primarily for differentiation between the biogenic amines and ascorbic acid). Modification schemes employed include carbon paste electrodes impregnated with stearic acid,[9] electrodes coated with charged Nafion films,[10] an enzyme electrode based on ascorbic acid oxidase,[11] and electrochemically treated electrodes.[5,12] It is hoped that new modification schemes will be developed for improving the stability of the implanted voltammetric sensors (which is hampered by contacting the biological fluid for extended periods of time).

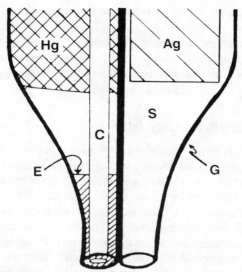

FIGURE 5-1
Schematic of a two-barrel microelectrode tip: (Ag) AgCl-coated Ag
wire; (C) carbon fiber; (E) epoxy; (G) glass capillary wall; (Hg)
mercury contact to the carbon fiber; (S) saline solution.
(Reproduced with permission.[7])

5-2.2 Measurement Schemes

Several waveforms have been used in conjunction with controlled-potential in-vivo
measurements. The most commonly used (particularly for neurochemical studies) is
the chronoamperometric technique. Standard voltammetric techniques yielding
peak signals and linear scan (or cyclic) voltammetry, as well as differential-pulse
voltammetry, have been used in various situations.[13, 14] The latter is useful for trace
analysis, but is characterized by relatively long measurement times. Theoretical
expressions for the response obtained by linear scan voltammetry or chronoamperometry when the electroactive substance is consumed from a restricted compartment
(as occurs upon implantation in the brain) have been presented.[15, 16] Normal-pulse
voltammetry at microvoltammetric electrodes has been shown to minimize electrode
poisoning, thus yielding stable voltammograms.[17] The theory of normal-pulse
voltammetry at microelectrodes was considered by Sujaritvanichpong et al.[18] Other
waveforms, such as differential-pulse amperometry[19] and differential–double-pulse
voltammetry,[20] can also be advantageous for various in-vivo applications.

Conventional three-electrode potentiostats are commonly employed for in-vivo
voltammetry. The reduced surface area of voltammetric microelectrodes often
requires some changes, particularly when monitoring low concentrations. These
include the use of high-quality current amplifiers (for current measurements in the
nanoamp or picoamp range) and increased shielding to minimize noise. Commercial
instruments designed especially for this purpose are available, including the

Bioanalytical Systems Model CV37 voltammetric analyzer and Tacussel's Biopulse voltammetric system.

Fairly sophisticated microcomputer-controlled systems have been designed for in-vivo voltammetric studies.[21] Such systems allow operation of several working electrodes simultaneously, measurement of low current levels, gain adjustment, and automation for long-term experiments.

5-2.3   Probing Brain Chemistry

Adams and coworkers at the University of Kansas pioneered the use of miniature carbon electrodes for in-vivo monitoring of neurotransmitters in small animal brains. Neurotransmitters have a very important role in the brain since they are the key link in the communication between the neurons. The ultimate goal of the electrochemical experiments is establishment of methodology that can be used reliably for the study of neurochemistry in-vivo. Most neurotransmitters are easily oxidized at carbon surfaces, and this forms the basis for their in-vivo detection. By using in-vivo electrochemistry, one could study changes in the levels of neurotransmitters during behavioral, pharmacological, or physiological manipulations of the animal. The challenges of this research have been reviewed by Adams.[22, 23] A practical experiment consists of recording the voltammetric response while the animal is receiving no purposeful stimulation. After establishing a stable baseline, a chemical or electrical stimulation is applied, and the changes in the voltammetric response are monitored as a function of time. For example, Figure 5-2 shows electrochemical measurements of amphetamine-induced neurotransmitter release. Measurements of this effect were performed with the working electrodes in the rat caudate nucleus (at both left and right caudates). Approximately 15 min after the amphetamine injection, the current rose by 20–40% and slowly returned to baseline within 1 h. While these data do not clearly indicate which compounds are changing in concentration, the method does respond to this pharmacological manipulation. Many similar studies, based on different stimulations, have been performed. For example, various behavioral effects in the animal (e.g., electrical or vibrational shocks) yield sharp signals related to the neurotransmitter release.[25] Figure 5-3 shows in-vivo chronoamperometric monitoring of homovanillic acid in cerebral spinal fluid (CSF). Electrical stimulation of substantia nigra is known to release homovanillic acid into the CSF. About 20 min after such stimulation, the homovanillic acid concentration rises sharply and then decays slowly by washout via the CSF flow. Ewing et al.[26] were able to measure electrically stimulated dopamine release on a time scale of seconds. Additional information may be obtained by coupling the voltammetric probes with ion-selective microelectrodes that provide concomitant information on ionic fluxes.[27]

While this research area is still exploratory, it appears to open up a whole new avenue in neurophysiology, based on chemical, rather than electrical sensing. The resulting studies have already yielded important information on the dynamics of neurotransmitter release and uptake, thus providing excellent insights into neurochemical activity in the living brain. The major problem facing in-vivo brain

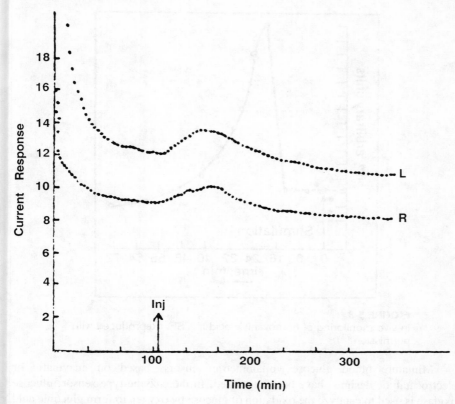

FIGURE 5-2
Amphetamine-induced dopamine release; electrochemical
monitoring in the left (L) and right (R) caudates. The animal
received a dose of 4 mg/kg of amphetamine. (Reproduced with
permission.[24])

studies is the lack of resolving power between the neurotransmitter of interest and
other oxidizable substances (particularly ascorbic acid). This problem can be
addressed successfully by using various modified microelectrodes [9-12] described
earlier. In particular, negatively-charged Nafion coatings exclude anionic interferences, such as ascorbic acid, while allowing facile detection of dopamine, which
contains a protonated amine group at physiological pH.[10, 28] To achieve the desired
rapid temporal response, a coating with minimal thickness is desired.[28]

5-2.4 In-Vivo Glucose Probes

Continuous on-line monitoring of blood glucose would provide better control of
diabetes, and ultimately may lead to internal insulin release systems for use in
diabetic treatment. All the components of such a feedback system have been made
except for a highly stable in-vivo glucose sensor.

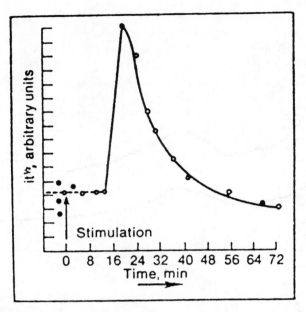

FIGURE 5-3
In-vivo monitoring of homovanillic acid in CSF. (Reproduced with permission.[22])

Miniature blood glucose voltammetric sensors, based on enzymatic or electrocatalytic designs, have been suggested. In the enzyme-type sensor, glucose oxidase is used to catalyze the oxidation of glucose by oxygen to form gluconic acid and hydrogen peroxide; the enzymatic reaction may be followed by electrochemical detection of either oxygen consumption or hydrogen peroxide production (Chapter 3). The advantage of this scheme is the specificity of glucose oxidase for glucose. Its major disadvantages for long-term in-vivo glucose sensing are the stoichiometric limitation of the enzymatic reaction by oxygen, the limited lifetime of the enzyme at body temperature, and interferences from other electroactive constituents. Glucose sensors based on this principle have been made in several configurations and have been employed in various glucose-monitoring situations.[29-35] Among these are a two-dimensional sensor design aimed at overcoming the oxygen deficit problem[35] and various needle glucose electrodes.[34] Although results are encouraging, stability of such sensors beyond several hours or days remains a problem. Thus far, the nearest approach to long-term monitoring has been that of Shichiri et al.,[32, 33] who reported successful in-vivo tissue monitoring of pancreatectomized dogs over periods of 3 days, and some shorter-term successes with human volunteers. A needle-type probe, consisting of a fine platinum wire coated with glucose oxidase and cellulose acetate and polyurethane membranes, was employed. An outer tube of silver-coated stainless steel served as the cathode. Modified electrodes, based on the attachment of various mediators or cofactors, e.g., ferrocene, into glucose oxidase–based in-vivo probes, hold great promise for minimizing interferences due to endogenous electroactive constituents or changes in the oxygen tension. These and other promising approaches to achieve oxygen-independent glucose probes (based primarily on a direct redox reaction of the enzyme) are discussed in Chapter 3.

The possibility of utilizing the surface of a metal as an electrochemical catalyst may offer a solution to some of the problems associated with the enzyme catalysis. Lerner et al.[36] investigated the utility of a platinum electrode for direct oxidation of glucose. Encouraging results were obtained in bovine serum. Similarly, Marincic et al.[37] obtained good reproducibility for the oxidation of glucose at platinum electrodes under simulated physiological conditions. Lewandowski et al.[38] demonstrated the utility of a platinum black electrode for short-term direct blood glucose measurements on dogs. Other metallic surfaces, such as gold and gold oxide, also appear promising as catalysts for glucose oxidation.[39] Various surface reactivation schemes are being explored to offer the desired long-term stability.

Diabetic patients can also benefit from the use of bedside monitors. For example, Miles Labs developed an electrode-based analyzer that provides continuous monitoring of glucose, as well as potassium and calcium.[40] Rapid measurements are obtained using small quantities of blood. To prevent clotting in the analyzer manifold, heparin is added to the sample line immediately after blood is withdrawn. Figure 5-4 illustrates the simultaneous on-line response for glucose and potassium in living dogs.

5-2.5   In-Vivo Oxygen Measurements

The determination of oxygen is one of the most important applications of electroanalysis in the clinical field. The reduction of oxygen on solid electrodes, held at a constant potential, is very useful for in-vivo monitoring of oxygen. The resulting current must be directly proportional to the partial pressure of oxygen in the analyzed medium. The platinum working electrode is usually covered by a gas-permeable hydrophobic membrane, which excludes potential interferences. The choice of the membrane and its thickness is important for obtaining reliable measurements. For example, thin membranes that offer fast response times exhibit strong dependence on oxygen mass transport in the analyzed medium. Other frequent problems are the temperature dependence of the current and the magnitude of the background current (especially when monitoring small amounts of oxygen). Catheter electrodes are usually employed for in-vivo measurements of oxygen in blood (e.g., Refs. 41 and 42), while various needle-type probes can be used for monitoring tissue oxygen pressure.[43] For example, Jank et al.[42] described the continuous recording of arterial oxygen pressure in humans. The oxygen catheter electrode was implanted between the radial artery and antecubital vein. Widespread use of invasive oxygen probes in humans would require greater attention to problems such as biocompatibility and sterilization. Animal experiments, in contrast, have been successful for routine clinical measurements for more than 25 years. Additional (noninvasive) means for in-vivo oxygen measurements based on transcutaneous probes are discussed in Section 5-4.

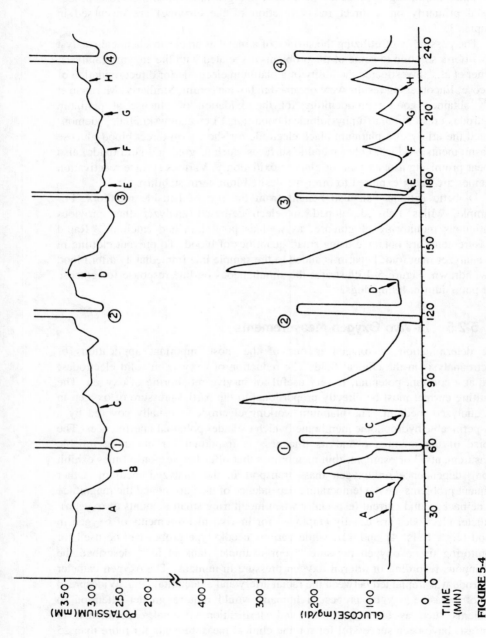

FIGURE 5-4

Concomitant on-line monitoring of glucose (bottom) and potassium (top). (1–4) Drifts checks with
standards; (A) injection of 1 unit of insulin; (B) injection of 4 g of dextrose; (C) injection of 8 g of dextrose;
(D) injection of 8 g of dextrose plus 2 units of insulin; (F, G) injection of 2 g of dextrose; (H) injection of

5-2.6   Other In-Vivo Voltammetric Applications

Voltammetric monitoring of ascorbic acid was used by Koryta and coworkers[44] for evaluation of the conditions of a kidney used for transplantation. Cyclic voltammograms were recorded in the cortex of the kidney using a platinum working electrode. The response to the injection and elution of ascorbic acid allowed discrimination between healthy kidneys (with a well-perfused cortex) and damaged kidneys (with low perfusion rates).

A long microelectrode assembly with a carbon paste working electrode was used by Wang and coworkers[14] for in-vivo monitoring in primate species (Figure 5-5). Acetaminophen was detected in rhesus monkey bloodstream following its intravenous administration. A rapid assessment of the blood concentration was reported. Morgan and Freed[45] illustrated that acetaminophen can be used as an internal standard for calibrating in-vivo voltammetric electrodes, and hence for minimizing problems of changes in sensitivity during implantation. Meulemans[46] described recently an indwelling carbon-rod catheter electrode for the in-vivo assay of electroactive antibiotic drugs in the bloodstream of rats. The electrode was introduced in the common carotid artery near the aortic arch (Figure 5-6). Differential-pulse voltammetry was used to monitor chloramphenicol and cefsulodin following their intravenous administration. Feher et al.[47] built a miniature flow cell with a silicone rubber-based graphite working electrode into the vein or artery of narcotized animals. A typical current-time curve recorded in a cat artery after the administration of amidopyrine is shown in Figure 5-7. An in-vivo life span of up to 4 h was reported. Future studies will undoubtedly include similar voltammetric probes responsive to other organic drugs. Routine operation of such probes in the clinical field will require substantial minimization of biological matrix interferences (adsorbing proteins, endogenous electroactive constituents, etc.). For example, coverage with an appropriate protective membrane is expected to yield improved stability, and hence to discriminate between surface fouling and drug metabolic effects (responsible to changes in the peak height). Improved selectivity is expected by immobilizing enzymes onto the in-vivo probe. The prospects in this direction have been reviewed.[48]

5-3   MINIATURE ION-SELECTIVE ELECTRODES FOR IN-VIVO MONITORING AND ASSAYS OF MICROLITER SAMPLES

5-3.1   Introduction

Efforts to use in-vivo monitoring to analyze important blood electrolytes have prompted considerable research into the design of miniaturized ion-selective electrodes. In-vivo ion-selective electrodes have many exciting possibilities for medical science. Walker[49] introduced ion-selective microelectrodes for intracellular measurements of potassium or chloride ions. Their design, shown in Figure 5-8, is based on a glass micropipet, with a tip drawn out to microscopic dimensions. Inside

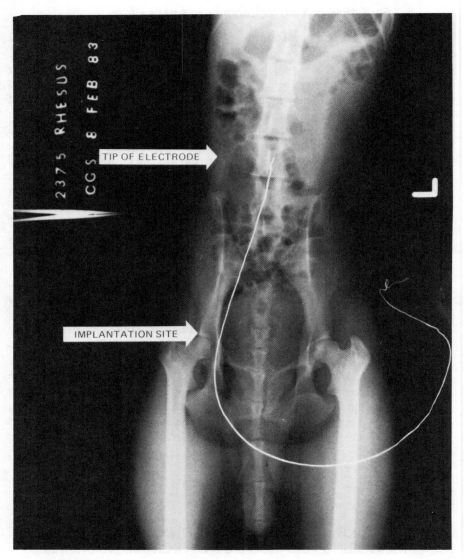

FIGURE 5-5
X-ray photograph showing the position of a voltammetric
microelectrode in the bloodstream of a rhesus monkey.
(Reproduced with permission.[14])

this tip, a droplet of liquid ion-exchanger is held by capillary action. The edges of
the tip are smoothed to minimize cell damage. A more recent approach to reduce the
size of conventional ion-selective electrodes is to place the ion-conducting
membrane material in direct contact with the electronic conductor, and hence
eliminating the internal reference solution. The electrode conductor may be a wire

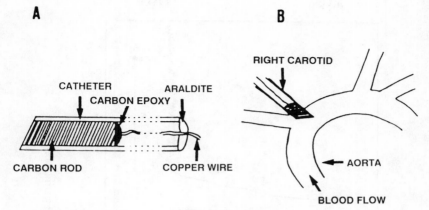

FIGURE 5-6
Catheter electrode for monitoring antibiotic drugs: (A) electrode
design; (B) implantation in the bloodstream. (Reproduced with
permission.[46])

or disk made of metal or graphite (as in the case of coated-wire electrodes) or an
insulating gate of a semiconductor (as in the case of ion-selective field effect
transistors). Prospects and problems involved in the operation of these coated
solid-state devices have been reviewed.[50]

One of the major difficulties in making reliable in-vivo potentiometric
measurements is selecting a stable reference electrode. The liquid junction has to be
highly stable, considering the fact that the total potential change over the normal
range of various electrolytes is only a few millivolts. Because of liquid junction
fluctuations and streaming potentials, many of the reference electrodes are
inappropriate for in-vivo sensing. The requirements for a stable in-vivo reference
electrode have been reviewed.[51] The reference should be as close as possible to the
sensing electrode, with no intervening structure between the two. Margules et al.[52]
developed a reference electrode specifically designed for in-vivo work. The
electrode exhibited high stability for 8 h during animal trials. Another difficulty
with in-vivo measurements is the interaction of certain blood components with the
sensing electrode.

For example, deposition of fibrin on a pH electrode has been shown to
substantially affect the response time.[53] The adsorption of biocomponents on the
test electrode may be responsible also for gradual changes in the measured potential.
Similar changes in the response may occur also due to a time-dependent electrical
field in excitable tissues. Details of construction and applications of ion-selective
microelectrodes have been reviewed.[44, 54, 55] Specific examples are described in the
following sections.

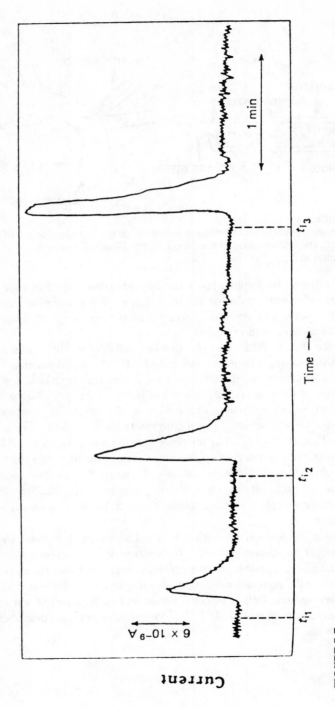

FIGURE 5-7
Current-time curves recorded in an artery after administration of amidopyrine. (Reproduced with permission.[47])

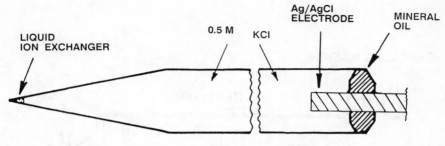

FIGURE 5-8
An ion-selective microelectrode. (Reproduced with permission.[49])

5-3.2 Monitoring of Potassium and Sodium Ions

Rapid analysis of potassium and sodium has become important for patient management situations. At the forefront of the research on continuous potentiometric monitoring of potassium and sodium is Wilhelm Simon of the Swiss Federal Institute of Technology. In an early study[56] he described a flow-through system with a valinomycin-based electrode (Figure 5-9) for the continuous on-line measurement of blood potassium ion concentration during human open-heart surgery. To insure normal heart function, the potassium level must be within the physiological range before reactivation of the heart. The electrochemical analyzer offered short (2-min) time delays, with possible reduction to 10–20 s. The results agreed well with those obtained with a flame photometer. (The latter method, however, requires centrifugation of the sample, and hence about 10-min delays.) The near-instantaneous electrochemical response promises to deepen insight and simplify rational treatment under surgery conditions. Simon's group later described microprocessor-controlled ex-vivo monitoring of sodium and potassium ion concentrations in undiluted urine of catheterized patients.[57] An automatic calibration system and an iterative procedure for the evaluation of activity coefficients were used for calculating ion concentrations. Successful operation in intensive-care units and during open-heart surgery was reported. Figure 5-10, for example, illustrates a fully automated recording during a heart operation for valve replacement (along with values obtained by flame photometry). The above development indicates great potential for routine medical applications.

Simon and coworkers[58] also developed a neutral carrier sodium-selective electrode, with a 1-μm diameter, for intracellular studies. Other groups have reported measurement of potassium ion activity in kidney proximal tubule,[59] central nervous system,[25] skeletal muscle,[60] and veins of a greyhound.[61] A catheter-type coated-wire probe, consisting of valinomycin in a silicon polymer matrix coated onto a silver wire, was used for in-vivo monitoring of potassium in venous blood.[62]

5-3.3 In-Vivo pH Measurements

Micro-pH electrodes are of essential importance to clinical monitoring and physiological studies. Early designs based on microbulb glass electrodes, assembled inside 23–26-gauge hypodermic needles, were described by Czaban and

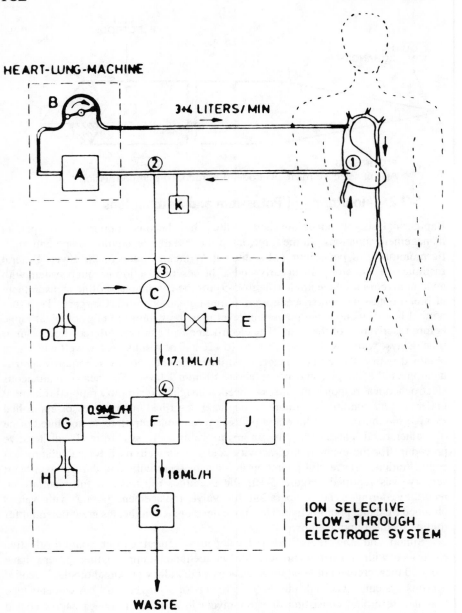

HEART-LUNG-MACHINE

3-4 LITERS/MIN

ION SELECTIVE FLOW-THROUGH ELECTRODE SYSTEM

WASTE

FIGURE 5-9
Flow-through system for on-line monitoring of blood potassium
concentration. (A) oxygenator; (B) pump; (C) sample-selection
valve; (D) standard solution; (E) air pump; (F) flow-through
electrode; (G) peristaltic pump; (H) reference electrolyte; (J) signal
acquisition and data output; (K) sampling syringe for off-line
measurements. (Reproduced with permission.[56])

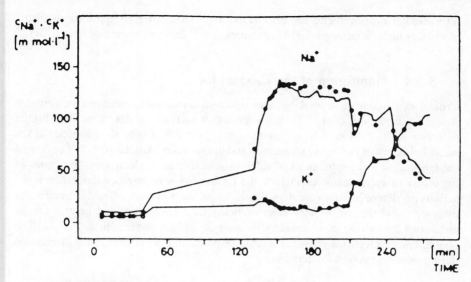

FIGURE 5-10
Automatic on-line determination of K$^+$ and Na$^+$ concentrations in urine during open-heart surgery. The symbols ● mark flame photometric determinations. (Reproduced with permission.[57])

Rechnitz.[63] Excellent response characteristics over a wide pH range were found. However, glass microelectrodes suffer from disadvantages associated with the tip geometry. Two factors limit the minimum size of glass membrane pH microelectrodes: the thickness of the pH glass and the configuration of the tip of the beveled pipet. Pucacco and Carter[64] described a method for thinning the pH glass to obtain submicrometer pH electrodes with rapid response times. Savinell et al.[65] described a miniature glass pH electrode with nonaqueous internal reference solution, which minimizes the tip gelling and consequently becomes more suitable for micropuncture applications. Stamm et al.[66] introduced a miniature glass electrode for continuous monitoring of tissue pH, particularly for use in the high-risk fetus and critically ill neonate. Glass electrodes also have high electrical impedances, which cause problems when trying to carry out reliable measurements in electrically noisy environments (e.g., intensive-care units, operating theaters). Nevertheless, glass membrane pH microelectrodes have been successfully applied in various clinical situations, e.g., for in-vivo measurements of changes in blood pH due to hyperventillation[67] and for continuous monitoring of the scalp tissue pH of the human fetus during labor.[68] Nonglass micro-pH electrodes, based on micro-liquid membranes (with appropriate ion-exchanger or neutral carrier),[69, 70] metal–metal oxide based on antimony,[71] palladium,[72] and tungsten,[73] and ion-selective field effect transistors[74] have all been considered as miniaturized pH probes. There even exists a commercially available electrode for continuous measurements of tissue pH.[75] Commercial catheter-tip pH ion-selective field effect transistors for use in intravascular measurements are also available.[76] Other in-vivo applications of pH electrodes include intracellular pH measurements[77] and monitoring of muscle pH.[78]

5-3.4   Monitoring of the Calcium Ion

Critical evaluation of several calcium-selective liquid membrane microelectrodes was given by Lanter et al.[79] They observed that such electrodes often yield higher detection limits than their larger counterparts, probably due to slow kinetics at the phase boundary of the liquid membrane and the solution. Ammann et al.[80] reviewed the fabrication and applications of calcium-selective microelectrodes. Examples of the utility of such electrodes include the measurement of intracellular calcium ion activity,[81] determination of calcium activity in sacroplasms of rabbit ventricular muscle,[82] and the use of ion-selective field effect transistors and neutral carrier electrodes for in-vivo continuous monitoring of ionized calcium in dogs.[83-85] The neutral carrier electrode provided a rapid response in the extracorporeal circulation during administration of calcitonin.

5-3.5   Other In-Vivo Potentiometric Probes

A number of other ion-selective microelectrodes suitable for assays of microliter samples or in-vivo monitoring have been described. Simon's group[86] described a neutral carrier-based microelectrode (1-μm tip diameter) for intracellular magnesium activity studies. Czaban and Rechnitz[87] developed solid-state ion-selective microelectrodes for measuring heavy metals and halides. A solid-state fluoride electrode for assays of nanoliter-size samples was described by Vogel et al.[88] Leader[89] described an improved solid-state silver–silver chloride microelectrode for intracellular measurements of chloride ion. Similarly, Aickin and Brading[90] used, over considerable periods, a chloride-selective microelectrode for monitoring chloride activity in smooth muscle cells of guinea pig vas deferens.

Frequent analysis of blood carbon dioxide is crucial to the management of surgical and intensive-care patients. Reliable in-vivo carbon dioxide probes are thus desirable for detecting serious changes in the level of blood gas. Opdycke and Meyerhoff[91] recently described a potentiometric P_{CO_2} sensing catheter based on an internal tubular polymer pH electrode and an outer gas-permeable silicone rubber tube (Figure 5-11). Such a configuration protects the sensing region from damage during catheter placement or removal, and allows significant size reduction. Successful results for sensors implanted intravascularly in a dog demonstrated the suitability for continuous in-vivo monitoring. In-vitro blood pump studies indicated good correlation with conventional blood gas instruments (Figure 5-12). Other miniature P_{CO_2} probes, which rely on fragile glass[92] or metal–metal oxide[93] pH electrodes, appear to be less suitable for continuous in-vivo monitoring.

Justice and coworkers[94] developed an ion-selective microelectrode for acetylcholine and choline based on their complexes with dipicrylamine. A near-Nernstian response was obtained over the 5×10^{-5}- to $10^{-2}M$ range. Such electrodes are

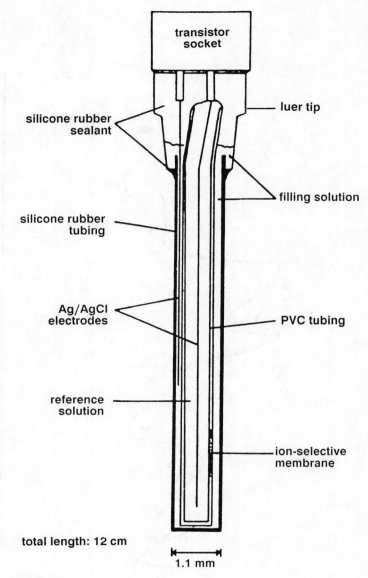

FIGURE 5-11
Schematic of a carbon dioxide—sensing catheter. (Reproduced with permission.[91])

suitable for in-vivo monitoring in the brain and should supplement voltammetric brain probing of easily oxidizable constituents.

Patients undergoing renal dialysis therapy can also benefit from the real-time capability of potentiometric sensors. For example, Klein and Whalthen[95] developed

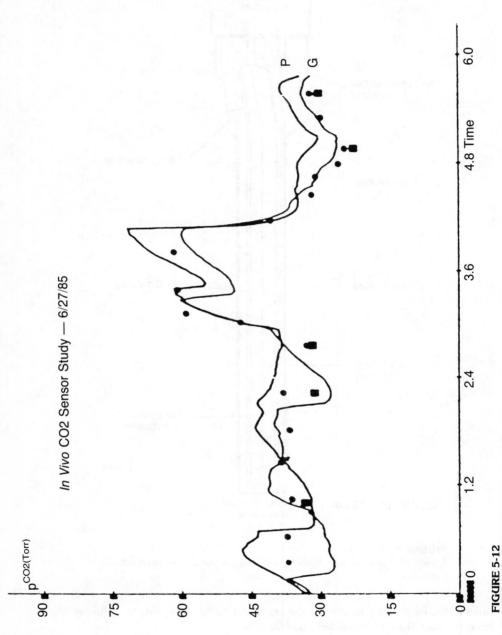

FIGURE 5-12
Continuous in-vivo monitoring of carbon dioxide using two catheters as compared with in-vitro blood gas

an instrument for continuous monitoring of urea in dialysate effluent from a dialyzer unit. A flow-through urease reactor was used to convert the urea into ammonia, which was detected downstream using an ammonium ion-selective electrode. Urea can be determined also using an enzymatic microsensor with a tip diameter of 10 μm.[96] Such a probe was fabricated by immobilizing urease at the tip of an air-gap ammonia microelectrode, and responded to 10^{-2}–10^{-4} M urea in 30–45 s.

5-4   CONTINUOUS MONITORING USING NONINVASIVE ELECTROCHEMICAL SENSORS

In order to avoid the hazards of some invasive sensors, noninvasive electrochemical probes have been developed recently, particularly for gas measurements. Transcutaneous or "through-the-skin" monitoring of oxygen or carbon dioxide has been popular in the last 5 years. Such measurements offer a tool for continuous monitoring of arterial P_{O_2} or P_{CO_2} without the need for repetitive sampling. This is particularly important in premature infants, where invasive sampling can lead to vascular and infection complications, waste of blood, and trauma. The transcutaneous measurement is based on the correlation between the gas tension and transcutaneous tension. To achieve good correlation, it is important to hyperfuse the skin. Commercial instruments are used to warm the skin to about 44°C, thus promoting the diffusion of capillary blood gases to the surface. The transcutaneous oxygen electrodes that are now used are, in essence, Clark electrodes that have been redesigned and miniaturized. Currently popular transcutaneous carbon dioxide probes are similar to common carbon dioxide gas sensors with either a microporous Teflon or silicon rubber membrane separating the skin surface from a bicarbonate internal electrolyte layer and a combination glass pH electrode. Future electrodes for transcutaneous measurements will undoubtedly offer simultaneous monitoring of P_{O_2}, P_{CO_2}, and pH.

Cassady[97] discussed several factors essential to achieve good correlation between the gas tension and the transcutaneous readings. These include proper sensor design and application, maximum hyperemia, appropriate comparisons, systems permeable to the gas of interest, and no interfering agents. To date, transcutaneous oxygen monitoring in neonates has been very useful and is the most widely accepted method. Transcutaneous measurements in adults are still the subject of much controversy, because of unsatisfactory correlation with data obtained in conventional blood gas analysis. Wimberly et al.[98] evaluated the use of transcutaneous carbon dioxide and oxygen probes in seven healthy adults. Reliable and rapid response was observed for carbon dioxide, with good correlation with capillary P_{CO_2} levels. The conclusions regarding transcutaneous oxygen values were less definite because of uncertain interpretation of the capillary P_{O_2} values. Because of drifts in the response, the intervals between calibrations should be limited to a few hours.[99]

Elevated chloride levels in sweat are one characteristic associated with cystic fibrosis. Warwick and Hansen[100] described the use of a skin-chloride-selective electrode for direct potentiometric measurement of chloride in sweat. The amount of pressure between the skin and the electrode and the contact time were shown to have little effect on the results. This flat-headed electrode can be used for rapid screening of babies in hospitals. Green et al.[101] described a potentiometric method for rapid sequential measurement of the activity of sodium, chloride, and hydrogen ions in sweat. Good precision was obtained, and factors such as temperature control and electrode-skin contact were discussed.

New noninvasive electrochemical sensors are expected to appear in the near future. For example, Guilbault's group is currently developing a probe for glucose measurements in the trans buccal mucosa.[102] Noninvasive probes for pH and bilirubin are being tested in other laboratories.[103] New fabrication technologies, such as thin-film printing, offer the possibility of low-cost disposable skin-surface sensors. The noninvasive strategy appears to be particularly advantageous in children for whom invasive monitoring can be traumatic. For more information on present activity and future prospects of this concept, the reader is referred to a recent review.[103]

5-5   CONCLUSION

The development of miniaturized potentiometric and voltammetric probes continues to be a "hot" area of research. Even though significant progress has been made, in-vivo monitoring of important clinical variables is still mostly experimental. If methodologies become sufficiently reliable, they could be extremely useful in numerous clinical situations (particularly, for a close surveillance of patients). Several transducers have already made important contributions to patient management in the clinical practice. Before testing new probes in humans, extensive animal testing is necessary. Future work on in-vivo electrochemistry will undoubtedly focus on the development of selective probes for organic drugs or additional electrolytes, the coupling of enzymatic and immunochemical reactions with in-vivo electrodes, the design of new noninvasive and multispecies probes, and the search for means of protecting in-vivo electrodes against matrix effects. A very promising concept is the use of a modified electrode that can deliver (release) drugs or other chemicals to a specific location at a given time in response to monitoring at a second electrode. With these and other developments and improvements, in-vivo electrodes will prove highly useful in the clinical field.

REFERENCES

1. Pinkerton, T. C., Lawson, B. L., *Clin. Chem.*, *28*, 1946 (1982).
2. Conti, J. C., Strope, E., Adams, R. N., Marsden, C. A., *Life Sci.*, *23*, 2705 (1978).
3. Wang, J., *Anal. Chem.*, *53*, 2280 (1981).
4. Wightman, R. M., *Anal. Chem.*, *53*, 1126A (1981).
5. Gonon, F. G., Fombarlet, C. M., Buda, M. J., Pujol, J. F., *Anal. Chem.*, *53*, 1386 (1981).

6. Edmonds, T. E., *Anal. Chim. Acta, 175,* 1 (1985).
7. Dayton, M. A., Ewing, A. G., Wightman, R. M., *Anal. Chem., 52,* 2392 (1980).
8. Meulemans, A., Poulain, B., Baux, G., Tauc, L., Henzel, D., *Anal. Chem., 58,* 2091 (1986).
9. Blaha, C. D., Lane, R. F., *Brain Res. Bull., 10,* 861 (1983).
10. Gerhardt, G. A., Oke, A. F., Nagy, G., Maghaddam, B., Adams, R. N., *Brain Res., 290,* 390 (1984).
11. Schenk, J. O., Miller, E., Adams, R. N., *Anal. Chem., 54,* 1453 (1982).
12. Falat, L., Cheng, H. Y., *Anal. Chem., 54,* 2108 (1982).
13. Wightman, M., Strope, E., Plotsky, P., Adams, R. N., *Brain Res., 159,* 55 (1978).
14. Wang, J., Hutchins, L. D., Selim, S., Cummins, L. B., *Bioelectrochem. Bioenerg., 12,* 193 (1984).
15. Albery, W. J., Goddard, N. J., Beck, T. W., Fillenez, M., O'Neill, R. D., *J. Electroanal. Chem., 161,* 221 (1984).
16. Cheng, H. Y., *J. Electroanal. Chem., 135,* 145 (1982).
17. Ewing, A. G., Dayton, M. A., Wightman, R. M., *Anal. Chem., 53,* 1842 (1981).
18. Sujaritvanichpong, S., Aoki, K., Tokuda, K., Matsuda, H., *J. Electroanal. Chem., 199,* 271 (1986).
19. Marcenac, F., Gonon, F., *Anal. Chem., 57,* 1778 (1985).
20. Lane, R., Hubbard, A. T., *Anal. Chem., 48,* 1287 (1976).
21. Cheng, H. Y., White, W., Adams, R. N., *Anal. Chem., 52,* 2445 (1980).
22. Adams, R. N., *Anal. Chem., 48,* 1126A (1976).
23. Adams, R. N., *Trends Neurosciences, 160,* 1 (1978).
24. Conti, J. C., Strope, E., Adams, R. N., Mardsen, C. A., *Life Sci., 23,* 2705 (1978).
25. Schenk, J. O., Miller, E., Adams, R. N., *J. Chem. Ed., 60,* 311 (1983).
26. Ewing, A. G., Bigelow, J. C., Wightman, R. M., *Science, 221,* 169 (1983).
27. Nagy, G., Moghaddam, B., Oke, A., Adams, R. N., *Neuroscience Lett., 55,* 119 (1985).
28. Kristensen, E. W., Kuhr, W. G., Wightman, R. M., *Anal. Chem., 59,* 1752 (1987).
29. Gough, D. A., Leypoldt, J. K., Armour, J. C., *Diabetes Care, 5,* 190 (1982).
30. Clark, L. C., Jr., Duggan, C. A., *Diabetes Care, 5,* 174 (1982).
31. Fisher, U., Abel, P., *Trans. Am. Soc. Artif. Intern. Organs, 28,* 245 (1982).
32. Shichiri, M., Kawamori, R., Yamasaki, Y., Hakui, N., Abe, H., *Lancet,* 1129 (1982).
33. Shichiri, M., Kawamori, R., Haku, N., Yamasaki, Y., Abe, H., *Diabetes, 33,* 1200 (1984).
34. Churchouse, S. J., Mullen, W. H., Keedy, F. H., Battersby, C. M., Vadgama, P. M., *Anal. Proceed., 23,* 146 (1986).
35. Gough, D. A., Lucisano, J. Y., Tse, P. H. S., *Anal. Chem., 57,* 2351 (1985).
36. Lerner, H., Giner, J., Soeldner, J. S., Colton, C. K., *Ann. NY Acad. Sci., 428,* 263 (1984).
37. Marincic, L., Soeldner, J. S., Giner, J., Morris, S. J., *Electrochem. Soc., 126,* 43 (1979).
38. Lewandowski, J. J., Szczepanska-Sadowska, E., Kirzymien, J., Nalecz, M., *Diabetes Care, 5,* 238 (1982).
39. Makovas, E. B., Liu, C. C., *Bioelectrochem. Bioenerg., 15,* 157 (1986).
40. Fogt, E. J., Eddy, A. R., Clemens, A. H., Fox, J., Heath, H., *Clin. Chem., 26,* 1425 (1980).
41. Yokota, H., Kreuzer, F., *Pflügers Arch. Ges. Physiol., 340,* 291 (1973).
42. Jank, K., de Hemptinne, J., Swietochowski, A., Demeester, M., *J. Appl. Physiol., 38,* 730 (1975).
43. Whalen, W. J., *J. Appl. Physiol., 23,* 798 (1967).
44. Koryta, J., Brezina, M., Pradac, J., Pradacova, J., in A. J. Bard (ed.), *Electroanalytical Chemistry,* Marcel Dekker, New York, 1979, vol. 11, p. 85.
45. Morgan, M. E., Freed, C. T., *J. Pharm. Exp. Ther., 219,* 49 (1981).

46. Meulemans, A., *Anal. Chem., 59*, 1872 (1987).
47. Feher, Z., Nagy, G., Toth, K., Pungor, E., *Analyst, 99*, 699 (1974).
48. Vadgama, P., *Trends Anal. Chem., 3*, 13 (1984).
49. Walker, J. L., *Anal. Chem., 43*, 89A (1971).
50. Arnold, M. A., Meyerhoff, M. E., *Anal. Chem., 56*, 20R (1984).
51. Tseung, A. C. A., Goffe, R. A., *Med. Biol. Eng. Comput., 16*, 677 (1978).
52. Margules, G. S., Hunter, G. M., MacGregor, D. C., *Med. Biol. Eng. Comput., 21*, 1 (1983).
53. Lofgren, O., *Arch. Gynecol., 226*, 17 (1978).
54. Somjen, G. G., in T. Zeuthen (ed.), *Applications of Ion-Selective Microelectrodes*, Elsevier, New York, 1981, chap. 11.
55. Krnjevic, K., Morris, M. E., in T. Zeuthen (ed.), *Applications of Ion-Selective Microelectrodes*, Elsevier, New York, 1981, chap. 12.
56. Osswald, H. F., Asper, R., Dimai, W., Simon, W., *Clin. Chem., 25*, 39 (1979).
57. Dutsch, S., Jenny, H. B., Schlatter, K. J., Perisset, P. M., Wolff, G., Clerc, J. T., Pretsch, E., Simon, W., *Anal. Chem., 57*, 578 (1985).
58. Steiner, R. A., Oehme, M., Ammann, D., Simon, W., *Anal. Chem., 51*, 351 (1979).
59. Khuri, R. N., Agulian, S. K., Wise, W. M., *Pluger's Arch., 39*, 1971 (1971).
60. Hnik, P., Vyskovil, F., Kriz, N., Holas, M., *Brain Res., 40*, 559 (1972).
61. Treasure, T., *Intens. Care Med., 4*, 83 (1978).
62. Hill, J. L., in D. W. Lubbers, H. Acker, and R. P. Buck (eds.), *Progress in Enzyme and Ion-Selective Electrodes*, Springer, Berlin, 1981, pp. 81–85.
63. Czaban, J. D., Rechnitz, G. A., *Anal. Chem., 48*, 277 (1976).
64. Pucacco, L. R., Carter, N. W., *Anal. Biochem., 89*, 151 (1987).
65. Savinell, R. F., Liu, C. C., Kowalsky, T. E., Puschett, J. B., *Anal. Chem., 153*, 552 (1981).
66. Stamm, O., Latscha, H., Janecek, P., *Am. J. Obstet. Gynecol., 124*, 193 (1976).
67. Band, D. M., Semple, S. J., *Proc. Physiol. Soc. (London)*, 58p. (1966).
68. Nickelsen, C., Thomsen, S. G., Weber, T., *Br. J. Obs. Gynec. 92*, 220 (1985).
69. Clare-Harmon, M., Pool-Wilson, P. A., *J. Physiol. (London), 315*, 1P (1981).
70. Ammann, D., Lanter, F., Steiner, R. A., Schulthess, P., Shijo, Y., Simon, W., *Anal. Chem., 53*, 2267 (1981).
71. Mackenzie, J. W., Salkind, A. J., Topaz, S. R., *J. Surg. Res., 16*, 632 (1974).
72. Grubb, W. T., King, L. H., *Anal. Chem., 52*, 270 (1980).
73. Caldwell, P. C., *J. Physiol. (London), 120*, 31p. (1953).
74. Bergveld, P., Bousse, L., *Ned. Tijdschr. Natuurkd. A., A49*, 74 (1983).
75. Mindt, W., Maurer, H., Moeller, W., *Arch. Gynecol., 226*, 9 (1978).
76. Bergveld, P., *Biosensors, 2*, 15 (1986).
77. Caldwell, P. C., *J. Physiol. (London), 126*, 169 (1954).
78. Gebert, G., Friedman, S. M., *J. Appl. Physiol., 34*, 122 (1973).
79. Lanter, F., Steiner, R. A., Ammann, D., Simon, W., *Anal. Chim. Acta, 135*, 51 (1982).
80. Ammann, D., Meier, P. C., Simon, W., in C. C. Ashley and A. K. Campbell (eds.), *Detection and Measurement of Free Calcium in Cells*, Elsevier, Amsterdam, 1979, pp. 117–129.
81. Hamaguchi, Y., *Role Calcium Biol. Syst., 1*, 85 (1982).
82. Lee, C. O., Uhm, D. Y., Dresdner, K., *Science, 209*, 699 (1980).
83. McKinley, B. A., Wong, K. C., Janata, J., *Crit. Care Med., 9*, 333 (1981).
84. Anker, P., Ammann, D., Meier, P. C., Simon, W., *Clin. Chem., 30*, 454 (1984).
85. McKinley, B. A., Saffle, J., Jordan, W. S., Janata, J., Moss, S. D., Westenskow, D. R., *Med. Instrument, 14*, 93 (1980).
86. Lanter, F., Erne, D., Ammann, D., Simon, W., *Anal. Chem., 52*, 2400 (1980).
87. Czaban, J. D., Rechnitz, G. A., *Anal. Chem., 45*, 471 (1973).
88. Vogel, G. L., Chow, L. C., Brown, W. E., *Anal. Chem., 52*, 375 (1980).

89. Leader, J. P., *Proc. Univ. Otago Med. Sch., 60,* 34 (1982).
90. Aickin, C. C., Brading, A. F., *J. Physiol., 326,* 139 (1982).
91. Opdycke, W., Meyerhoff, M. E., *Anal. Chem., 58,* 950 (1986).
92. Lai, N. C., Liu, C. C., Brown, E. G., Neuman, M. R., Ko, W. H., *Med. Biol. Eng., 13,* 876 (1975).
93. Van Kempem, L. H. J., Kreuzer, F., *Respir. Physiol., 23,* 371 (1975).
94. Jaramillo, A., Lopez, S., Justice, J. B., Salamone, J. D., Neill, D. B., *Anal. Chim. Acta., 146,* 149 (1983).
95. Klein, E., Whalthen, R. L., U.S. Patent 4,244,787 (1981).
96. Joseph, J. P., *Anal. Chim. Acta, 169,* 249 (1985).
97. Cassady, G., *J. Pediatr., 103,* 837 (1983).
98. Wimberly, P. D., Pederson, K. G., Thode, J., Fogh-Anderson, N., Sorensen, A. M., Siggaard-Andersen, O., *Clin. Chem., 29,* 1471 (1983).
99. Kost, G. J., Chow, J. L., Kenny, M. A., *Clin. Chem., 29,* 1534 (1983).
100. Warwick, W. J., Hansen, L. G., *Clin. Chem., 24,* 381 (1978).
101. Green, M., Behrendt, H., Libien, G., *Clin. Chem., 18,* 427 (1972).
102. Guilbault, G. G., International Symposium on Electroanalysis and Sensors, Cardiff, April 1987, paper 52.
103. Hicks, J. M., *Clin. Chem., 31,* 1931 (1985).

INDEX

INDEX

A

ac voltammetry, 13
acetaminophen, 36, 119, 131, 157
adenosine, 92
adriamycin, 35, 133
adsorptive stripping voltammetry, 19
albumin, 43
alcohol, 128
amidopyrine, 157
amino acids, 87
ammonia, 69, 138
anodic stripping voltammetry, 18
antibiotic agents, 34
anticancer drugs, 35, 133
antidepressants, 34
antimony, 31
arginine, 97
aromatic amines, 124
artificial enzymes, 93
ascorbic acid, 38, 96, 127, 138
automated flow systems, 135

B

background current, 9
benzodiazepine, 33
bicarbonate, 65
bienzyme sensor, 92
bilayer lipid membrane, 102
bilirubin, 41
bioaffinity sensor, 103
biosensors, 79
biotin, 103
blood, 29
brain analysis, 152
bromazepam, 33

C

cadmium, 28, 31
calcium, 60, 136, 164
carbohydrates, 128
carbonate, 65
carbon dioxide, 67, 137, 164, 167
cardiovascular drugs, 35, 133
catalytic waves, 41

catecholamines, 42, 122
cathodic stripping voltammetry, 18
cephalosporin, 98
chloramphenicol, 11, 35, 128, 157
chlordiazepoxide, 131
chloride, 64, 167
cholesterol, 90
chronoamperometry, 15, 152
cimetidine, 37
cis-dichlorodiammine-platinum, 35, 133
conductivity detection, 120
Cottrell equation, 16
coulometry, 24
creatinine, 89, 139
current, 2, 8
cyclic voltammetry, 20
cysteine, 41, 96
cytochrome C, 41

D

detectors, 114
diazepam, 34, 131
differential pulse voltammetry, 11
digoxin, 35, 133, 141
direct potentiometry, 53
disulfides, 125
DNA, 41
dopamine, 97, 122, 152
dry-reagent slides, 73
dual-electrode detection, 118

E

electroactive functionalities, 8, 110
electrodes
   air-gap, 66
   bacterial, 97
   carbon,
      carbon paste, 113
      glassy carbon, 113
   coated wire, 70
   enzyme, 81, 139
   glass pH, 54
   ion selective, 49, 157
   mercury, 3, 5

175